# Woodworking Tools

## 1600-1900

Peter C. Welsh

# Woodworking Tools 1600-1900

By

Peter C. Welsh

Contributions From
The Museum of History and Technology:
Paper 51

Cover Photograph: Steve-h

ISBN: 978-1-78139-159-4

# Contents

# List of Illustrations

# WOODWORKING TOOLS
## 1600–1900

*This history of woodworking hand tools from the 17th to the 20th century is one of a very gradual evolution of tools through generations of craftsmen. As a result, the sources of changes in design are almost impossible to ascertain. Published sources, moreover, have been concerned primarily with the object shaped by the tool rather than the tool itself. The resulting scarcity of information is somewhat compensated for by collections in museums and restorations.*

*In this paper, the author spans three centuries in discussing the specialization, configuration, and change of woodworking tools in the United States.*

THE AUTHOR: *Peter C. Welsh is curator, Growth of the United States, in the Smithsonian Institution's Museum of History and Technology.*

IN 1918, PROFESSOR W.M.F. PETRIE concluded a brief article on "History in Tools" with a reminder that the history of this subject "has yet to be studied," and lamented the survival of so few precisely dated specimens. What Petrie found so discouraging in studying the implements of the ancient world has consistently plagued those concerned with tools of more recent vintage. Anonymity is the chief characteristic of hand tools of the last three centuries. The reasons are many: first, the tool is an object of daily use, subjected while in service to hard wear and, in some cases, ultimate destruction; second, a tool's usefulness is apt to continue through many years and through the hands of several generations of craftsmen, with the result that its origins become lost; third, the achievement of an implement of demonstrated proficiency dictated against radical, and therefore easily datable, changes in shape or style; and fourth, dated survivals needed to establish a range of firm control specimens for the better identification of unknowns, particularly the wooden elements of tools—handles, moldings, and plane bodies—are frustratingly few in non-arid archaeological sites. When tracing the provenance of American tools there is the additional problem of heterogeneous origins and shapes—that is, what was the appearance of a given tool prior to its standardization in England and the United States? The answer requires a brief summary of the origin of selected tool shapes, particularly those whose form was common to both the British Isles and the Continent in the 17th century. Beyond this, when did the shape of English tools begin to differ from the shape of tools of the Continent? Finally, what tool forms predominated in American usage and when, if in fact ever, did any of these tools achieve a distinctly American character? In the process of framing answers to these questions, one is confronted by a constantly diminishing literature, coupled with a steadily increasing number of tool types.[1]

Figure 1.—1685: THE PRINCIPAL TOOLS that the carpenter needed to frame a house, as listed by JOHANN AMOS COMENIUS in his *Orbis Sensualium Pictus* were the felling axe (4), wedge and beetle (7 and 8), chip axe (10), saw (12), trestle (14), and pulley (15). (Charles Hoole transl., London, 1685. *Courtesy of the Folger Shakespeare Library.*)

---

[1] W.M. Flinders Petrie, "History in Tools," Annual Report Smithsonian Institution, 1918, pp. 563–572 [reprint].

Figure 2.—1685: THE BOXMAKER AND TURNER as pictured by Comenius required planes (3 and 5), workbench (4), auger (6), knife (7), and lathe (14). (From Johann Amos Comenius, *Orbis Sensualium Pictus.* *Courtesy of the Folger Shakespeare Library.*)

The literature of the subject, both new and old, is sparse, with interest always centering upon the object shaped by the craftsman's tool rather than upon the tool itself. Henry Mercer's *Ancient Carpenters' Tools*, first published in 1929, is an exception. It remains a rich source of information based primarily on the marvelous collections preserved by the Bucks County Historical Society. Since 1933, the Early American Industries Association, both through collecting and through its *Chronicle*, has called attention to the vanishing trades, their tools and techniques; the magazine *Antiques* has occasionally dealt with this subject. Historians of economic and industrial development usually neglect the tools of the woodcrafts, and when considering the toolmakers, they have reference only to the inventors and producers of machine tools. The dearth of written material is somewhat compensated for by the collections of hand tools in American museums and restorations, notably those at Williamsburg, Cooperstown, Old Sturbridge Village, Winterthur, the Henry Ford Museum, and Shelburne; at the latter in particular the extensive collection has been bolstered by

Frank H. Wildung's museum pamphlet, "Woodworking Tools at Shelburne Museum." The most informative recent American work on the subject is Eric Sloane's handsomely illustrated *A Museum of Early American Tools*, published in 1964. Going beyond just the tools of the woodworker, Sloane's book also includes agricultural implements. It is a delightful combination of appreciation of early design, nostalgia, and useful fact.

Charles Hummel's forthcoming *With Hammer in Hand: The Dominy Craftsmen of East Hampton*—to be published by the Yale University Press—will be a major contribution to the literature dealing with Anglo-American woodworking tools. Hummel's book will place in perspective Winterthur Museum's uniquely documented Dominy Woodshop Collection. This extensive collection of tools—over a thousand in number—is rich in attributed and dated examples which range from the early 18th through the mid-19th century. The literature of the subject has been greatly enhanced by the English writer, W.L. Goodman. Extending a series of articles that first appeared in the *Journal of The Institute of Handicraft Teachers*, Goodman has put together a well-researched *History of Woodworking Tools* (London, 1964), one particularly useful for its wealth of illustration from antiquity and the Middle Ages.

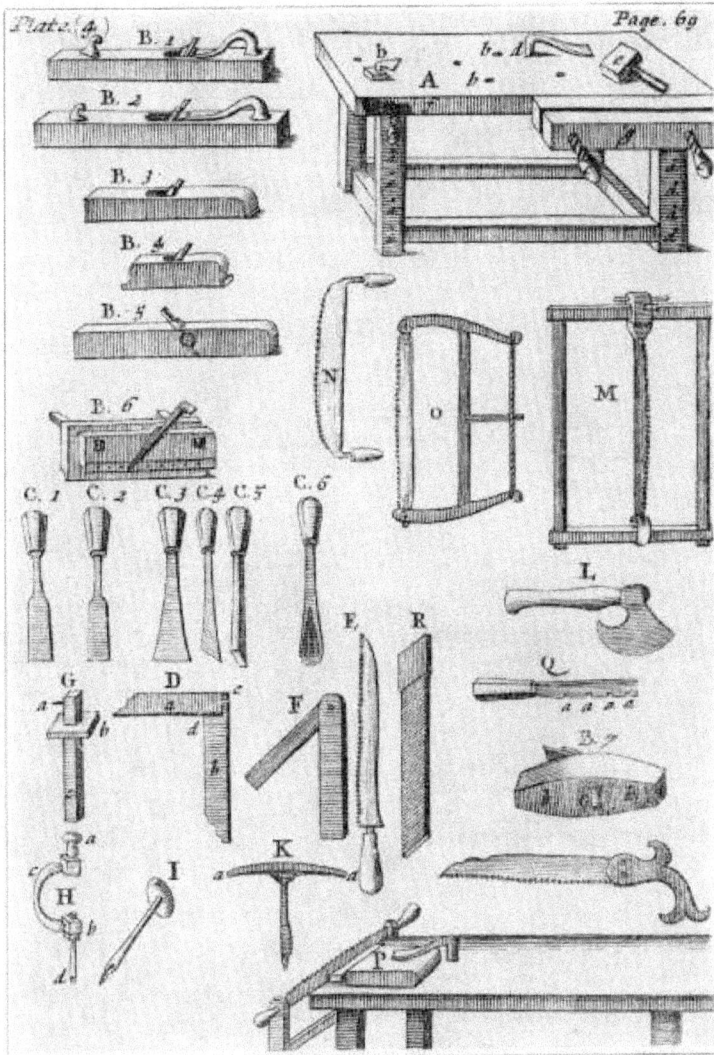

Figure 3.—1703: THE TOOLS OF THE JOINER illustrated by Moxon are the workbench (A), fore plane (B. 1), jointer (B. 2), strike-block (B. 3), smoothing plane (B. 4 and B. 7), rabbet plane (B. 5), plow (B. 6), forming chisels (C. 1 and C. 3), paring chisel (C. 2), skew former (C. 4), mortising chisel (sec. C. 5), gouge (C. 6), square (D), bevel (F), gauge (G), brace and bit (H), gimlet (I), auger (K), hatchet (L), pit saw (M), whipsaw (N), frame saw (O), saw set (Q), handsaw (unmarked), and compass saw (E). (Joseph Moxon, *Mechanick Exercises ...*, 3rd ed., London, 1703. Library of Congress.)

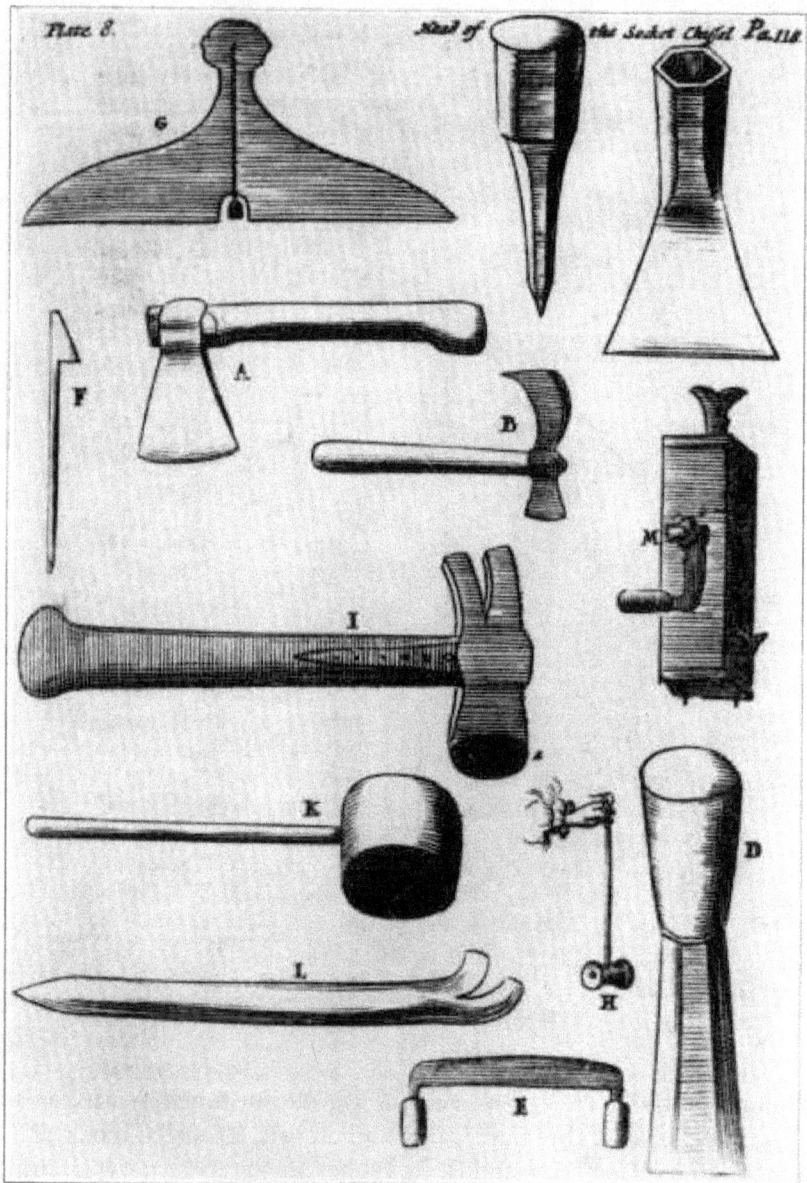

Figure 4.—1703: ONLY THE PRINCIPAL TOOLS used in carpentry are listed by Moxon: the axe (A), adz (B), socket chisel (C), ripping chisel (D), drawknife (E), hookpin (F), bevel (G), plumb line (H), hammer (I), commander (K), crow (L), and jack (M). (Moxon, *Mechanick Exercises ...*, 1703. Library of Congress.)

Peter C. Welsh

# Specialization

Given the limitations of precise dating, uncertain provenance, and an uneven literature, what can be learned about woodworking tools after 1600? In some instances, design change can be noted and documented to provide at least a general criteria for dating. Frequently, the original appearance of tools can be documented. For some hand tools, characteristics can be established that denote a national origin. Not infrequently a tool's style, decorative motif, or similarity to other objects that coexisted at a given time can suggest, even in relatively modern times, the values of the society that produced it. The source of such information derived from the hand tool is generally visual, recorded in the tool itself or in pictures of it and supported by manuscript and printed material.

Survey the principal printed sources of the 17th, 18th, and 19th centuries. The first thing that is apparent is a remarkable proliferation of tool types without any significant change in the definition and description of the carpenter's or joiner's task. Begin in 1685 with Charles Hoole's translation of Johann Amos Comenius' *Orbis Sensualium Pictus* for use as a Latin grammar. Among the occupations chosen to illustrate vocabulary and usage were the carpenter (fig. 1), the boxmaker (cabinetmaker), and the turner (fig. 2). "The Carpenter," according to Hoole's text, "squareth Timber with a Chip ax ... and saweth it with a Saw" while the more specialized "Box-maker, smootheth hewen-Boards with a Plain upon a Work-board, he maketh them very smooth with a little plain, he boarth them thorow with an Augre, carveth them with a Knife, fasteneth them together with Glew, and Cramp-irons, and maketh Tables, Boards, Chests &c." Hoole repeated Comenius' plates with the result that the craftsman's tools and his work have the same characteristic medieval flavor as the text.[1]

Joseph Moxon in his well-quoted work on the mechanic arts defined joinery as "an Art Manual, whereby several Pieces of Wood are so fitted and join'd together by Straight-line, Squares, Miters or any Bevel, that they shall seem one intire Piece." Including the workbench, Moxon described and illustrated 30 tools (fig. 3) needed by the joiner. The carpenter's tools were less favored by illustration; only 13 were pictured (fig. 4). The tools that the carpenter used were the same as

---

[1] Johann Amos Comenius, Orbis Sensualium Pictus, transl. Charles Hoole (London, 1685), pp. 130, 143.

those of the joiner except that the carpenter's tools were structurally stronger. The axe serves as a good example of the difference. The joiner's axe was light and short handled with the left side of the cutting edge bezeled to accommodate one-handed use. The carpenter's axe, on the other hand, was intended "to hew great Stuff" and was made deeper and heavier to facilitate the squaring and beveling of timbers.[1] By mid-18th century the craft of joiner and carpenter had been completely rationalized in Diderot's *Encyclopédie* and by André Roubo in his *L'Art du menuisier*, a part of Duhamel's *Descriptions des arts et métiers*. Diderot, for example, illustrates 14 bench planes alone, generally used by the joiner (fig. 5), while Roubo suggests the steady sophistication of the art in a plate showing the special planes and irons required for fine molding and paneling (fig. 6).

Despite such thoroughness, without the addition of the several plates it would be almost impossible to visualize, through the descriptive text alone, the work of the carpenter and joiner except, of course, in modern terms. This is particularly true of the numerous texts on building, such as Batty Langley's *The Builder's Complete Assistant* (1738) and Francis Price's *The British Carpenter* (1765), where building techniques are well described but illustration of tools is omitted. This inadequacy grows. In two 19th-century American editions of British works, *The Book of Trades*, printed at Philadelphia in 1807, and Hazen's *Panorama of the Professions and Trades* (1838), the descriptions of the carpenter's trade are extremely elementary.

Thomas Martin's *Circle of the Mechanical Arts* (1813), although far more thorough than many texts, still defined carpentry "as the art of cutting out, framing, and joining large pieces of wood, to be used in building" and joinery as "small work" or what "is called by the French, *menuiserie*." Martin enumerated 16 tools most useful to the carpenter and 21 commonly used by the joiner; in summary, he noted, as had Moxon, that "both these arts are subservient to architecture, being employed in raising, roofing, flooring and ornamenting buildings of all kinds" (fig. 7).[2]

In Peter Nicholson's *The Mechanic's Companion* (figs. 8, 9, and 10), the all-too-familiar definition of carpentry as "the art of employing timber in the construction of buildings" suggests very little of the

---

[1] Joseph Moxon, Mechanick Exercises or the Doctrine of Handy-Works, 3rd. ed. (London, 1703), pp. 63, 119.
[2] Martin, Circle of the Mechanical Arts (1813), p. 123.

carpenter's actual work or the improvement in tool design that had occurred since Moxon's *Exercises*. From Nicholson's list of the tools

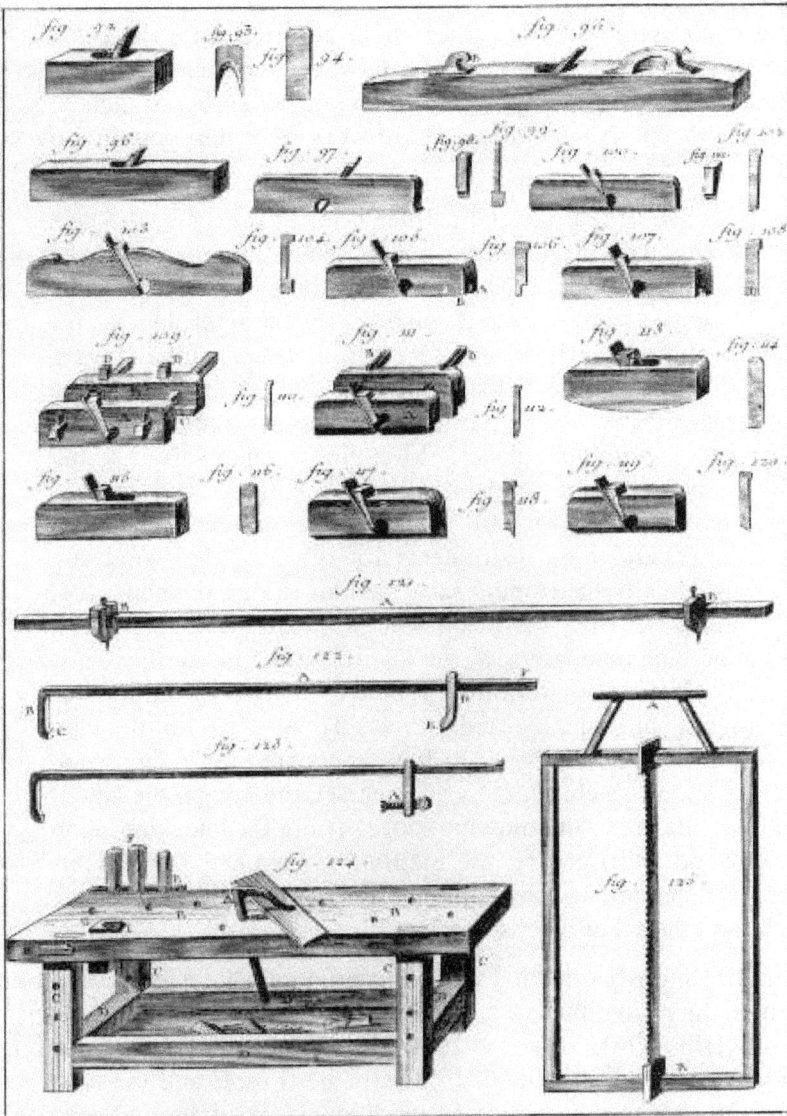

Figure 5.—1769: THE BENCH PLANES OF THE JOINER increased in number, but in appearance they remained much the same as those illustrated by Moxon. (Denis Diderot, *Recueil de planches sur les science et les arts libéraux*, Paris, 1769, vol. 7, "Menuiserie." Smithsonian photo 56630.)

required by the carpenter—"a ripping saw, a hand saw, an axe, an adze, a socket chisel, a firmer chisel, a ripping chisel, an auguer, a gimlet, a hammer, a mallet, a pair of pincers, and sometimes planes"—there would seem at first glance slight advance since the 1600's. The enumeration of the joiner's tools, however, indicates a considerable proliferation, particularly when compared to earlier writers. By the early 19th century, the more refined work of joinery required over 50 tools.

> The bench planes [instructed Nicholson] are, the jack plane, the fore plane, the trying plane, the long plane, the jointer, and the smoothing plane; the cylindric plane, the compass and forkstaff planes; the straight block, for straightening short edges. Rebating planes are the moving fillister, the sash fillister, the common rebating plane, the side rebating plane. Grooving planes are the plough and dado grooving planes. Moulding planes are sinking snipebills, side snipebills, beads, hollows and rounds, ovolos and ogees. Boring tools are: gimlets, bradawls, stock, and bits. Instruments for dividing the wood, are principally the ripping saw, the half ripper, the hand saw, the panel saw, the tenon saw, the carcase saw, the sash saw, the compass saw, the keyhole saw, and turning saw. Tools used for forming the angles of two adjoining surfaces, are squares and bevels. Tools used for drawing parallel lines are gauges. Edge tools are the firmer chisel, the mortise chisel, the socket chisel, the gouge, the hatchet, the adze, the drawing knife. Tools for knocking upon wood and iron are, the mallet and hammer. Implements for sharpening tools are the grinding stone, the rub stone, and the oil or whet stone.[1]

Reflecting what the text writers listed, toolmakers by the end of the 18th century gave buyers a wide choice. The catalogue of Sheffield's Castle Hill Works offered 20 combinations of ready-stocked tool chests; the simplest contained 12 carpenter's tools and the most complex, 39, plus, if desired, an additional assortment of gardening implements (fig. 11). In 1857, the Arrowmammett Works of Mid-

---

[1] Peter Nicholson, The Mechanic's Companion (Philadelphia, 1832), pp. 31, 89–90.

Peter C. Welsh

dletown, Connecticut, producers of bench and molding planes, published an illustrated catalogue that offered 34 distinct types that
included everything from hollows and rounds to double jointers and
hand-rail planes (fig. 12).[1]

---

[1] Catalog, Book 87, Cutler and Co., Castle Hill Works, Sheffield [in the collections of the Victoria and Albert Museum, London]; and Illustrated
Supplement to the Catalogue of Bench Planes, Arrowmammett Works (Middletown, Conn., 1857) [in the Smithsonian Institution Library].

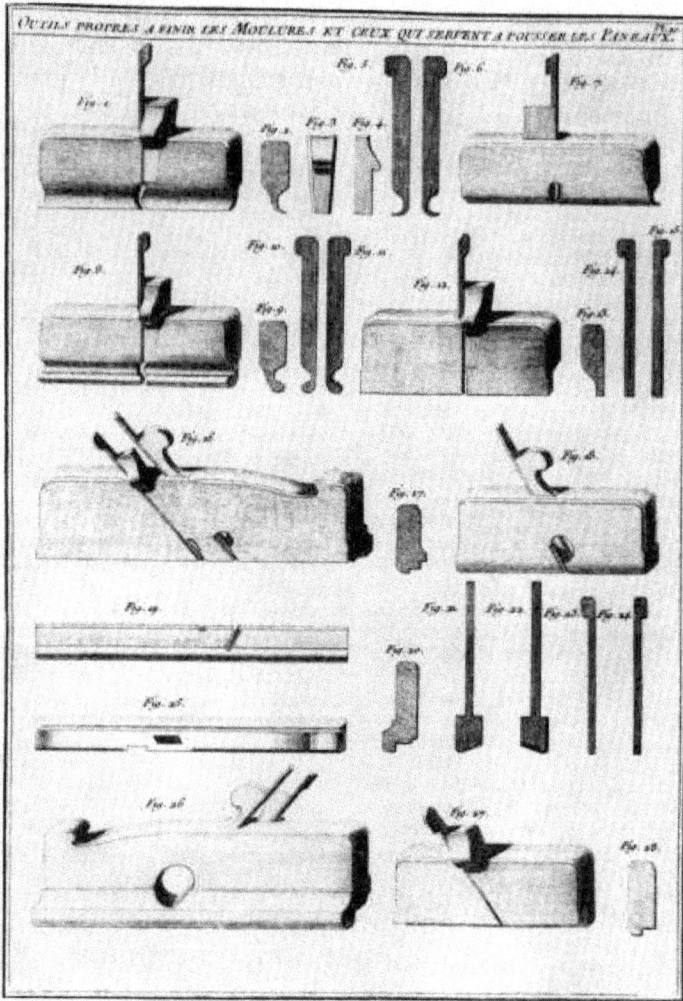

Figure 6.—1774: ANDRÉ ROUBO'S *L'Art du menuisier* contains detailed plates and descriptions of the most specialized of woodworking planes: those used to cut panel moldings. The conformation of these tools was still distinctly in keeping with the Moxon type and suggests that, at least in Europe, no remarkable change had yet occurred in the shape of planes. (André-Jacob Roubo, *L'Art du menuisier*: Troisième partie, troisième section, l'art du menuisier ébéniste [Paris, 1774]. Smithsonian photo 49790-D.)

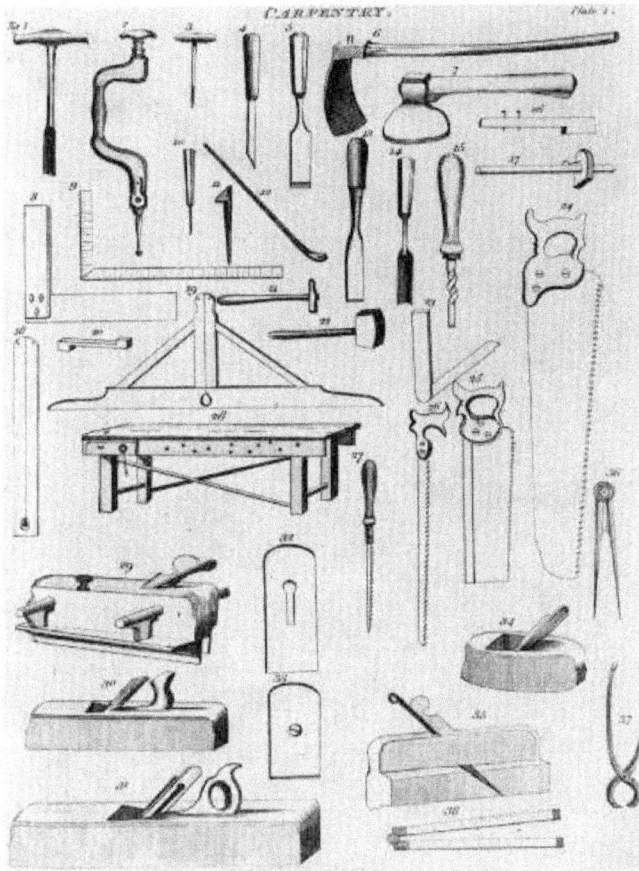

Figure 7.—1813: THOMAS MARTIN ILLUSTRATED ON ONE PLATE the tools of the carpenter and joiner dividing them as follows: the tools most useful to the carpenter, the axe (7), adz (6), saw (24), socket chisel (13), firmer chisel (5), auger (1), gimlet (3), gauge (16), square (9), compass (36), hammer (21), mallet (22), hookpin (11), crow (12), plumb rule (18), and level (19); and the tools most often associated with joinery, the jack plane (30), trying plane (31), smoothing plane (34), tenon saw (25), compass saw (26), keyhole saw (27), square (8), bevel (23), gauge (17), mortise chisel (4), gouge (14), turnscrew (15), plow plane (29), molding plane (35), pincers (37), bradawl (10), stock and bit (2), sidehook (20), workbench (28), and rule (38). The planes are of particular interest since they show clearly a change in form from those previously illustrated. (Thomas Martin, *The Circle of the Mechanical Arts*, London, 1813.)

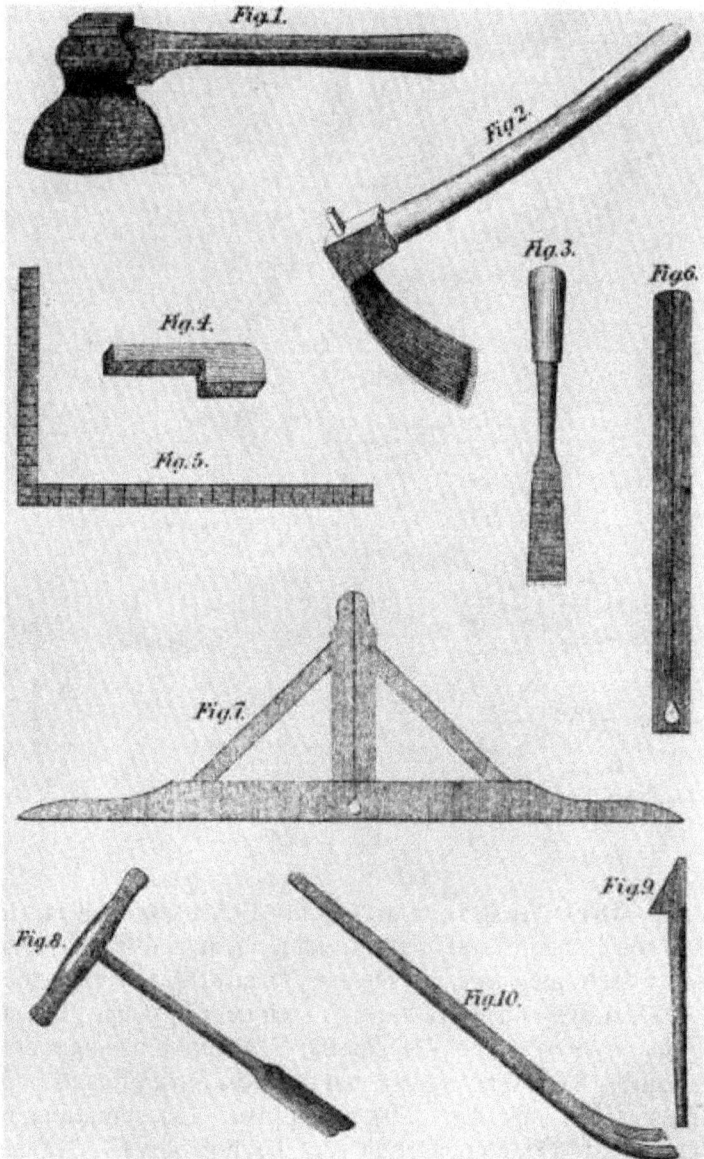

Figure 8.—1832: PETER NICHOLSON ILLUSTRATED an interesting mixture of old and new forms. An updating of Moxon, Nicholson's carpenter required an axe (1), adz (2), socket chisel (3), mortise and tenon gauge (4), square (5), plumb rule (6), level (7), auger (8), hookpin (9), and crow (10). (Peter Nicholson, *The Mechanic's Companion.* 1st American ed., Philadelphia, 1832. Smithsonian photo 56633.)

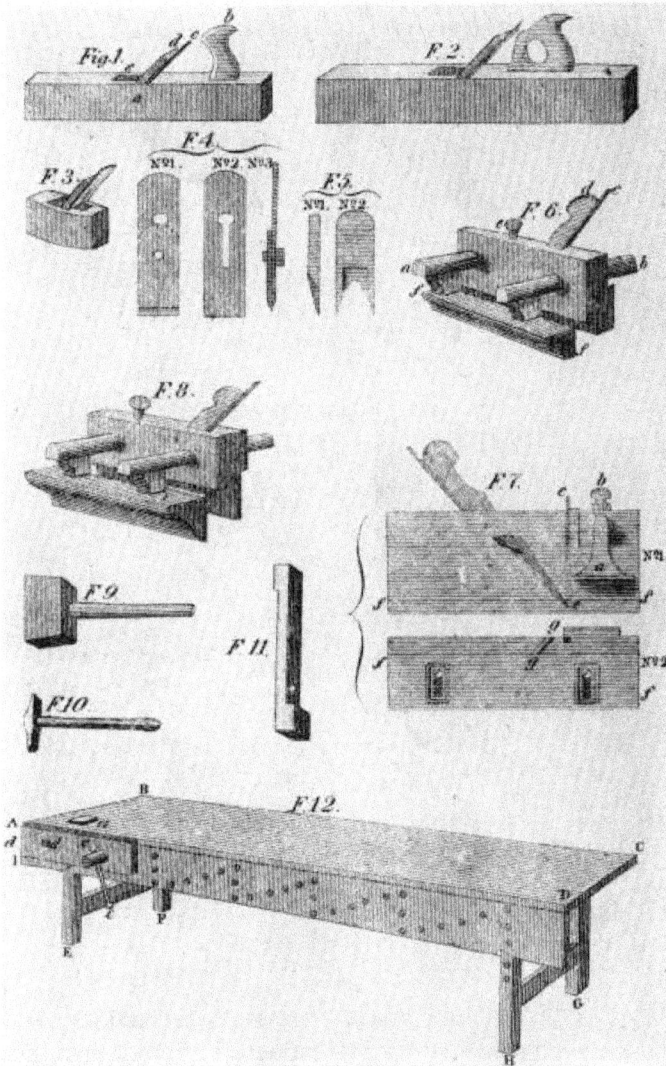

Figure 9.—1832: THE WORKBENCH DELINEATED BY NICHOLSON was little improved over Moxon's, although the planes—jack (1), trying plane (2), smoothing plane (3), sash fillister (7), and plow (8)—followed the form seen in Martin (fig. 7). The inception of this shape occurred in the shops of Sheffield toolmakers in the last half of the 18th century, and it persisted until replaced by metallic versions patented by American innovators during the last quarter of the 19th century. (Nicholson, *The Mechanic's Companion.* Smithsonian photo 56631.)

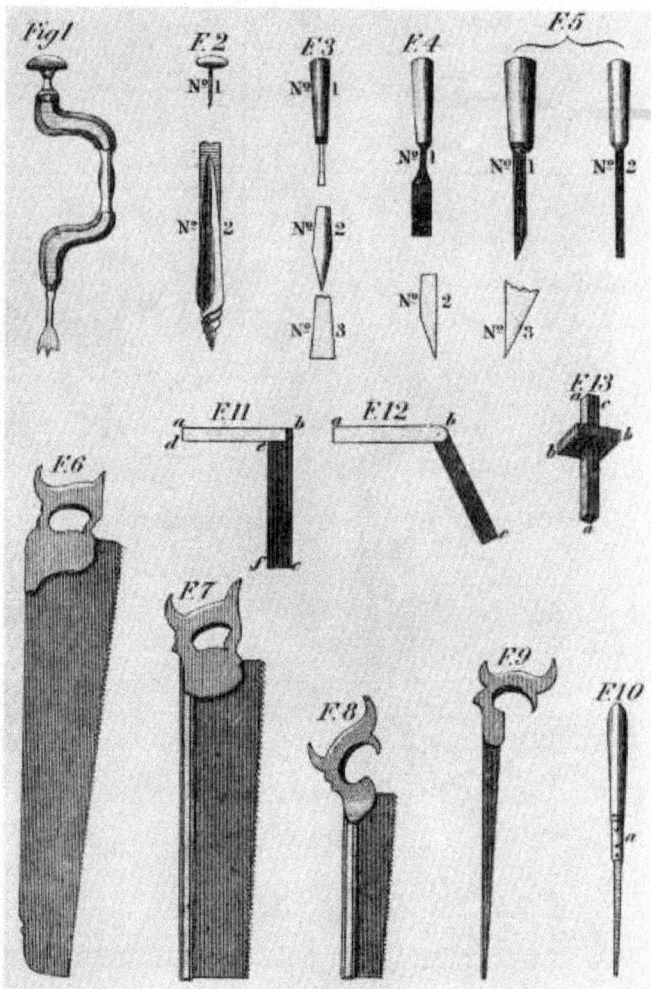

Figure 10.—1832: THE BRACE AND BIT, GIMLET, CHISELS, AND SAWS, having achieved a standard form distinctly different than those of Moxon's vintage, were, like the plane, slow to change. The metallic version of the brace did not replace the standard Sheffield type (1) in the United States until after 1850. For all intent and purpose the saw still retains the characteristics illustrated in Nicholson. Of interest is Nicholson's comment regarding the saws; namely, that the double handle was peculiar to the hand (6) and tenon saws (7), while the compass (9) and the sash saws (8) had the single handle. In addition the tenon saw was generally backed in iron and the sash saw in brass. (Nicholson, *The Mechanic's Companion*. Smithsonian photo 56632.)

## GENTLEMEN's Oak Tool Chests, *of different Sorts, as follow:*

Nº. 6, contains as under:—

In the chest is contained, a hammer, saw set, pair bright pincers, pair bright cutting nippers, a pair bright plyers, 1 hand vice, a foot rule, pair steel compasses, pair rack compasses, a striking knife, 1 brad punch, 4 brad awls, 6 gimblets sorted, 9 turn screws, a claw wrench, 1 bright chissel, and sundry partitions, containing a great variety of nails, wood screws, brass work, &c. &c. In a drawer is contained, a brace with 12 bits sorted, a small bick iron, a bevil and square, a spoke-shave, a line and roller, 5 files; 1 tulip and key-hole saw, 4 firmers, 2 gouges, and 3 mortice chissels. In another drawer is contained, an oil stone in a wood case, a mallet, jack plane, smoothing plane, bell hand saw, bell dove-tail saw, hatchet, and a glue pot.

Nº. 7, contains as under:—

In the chest and in two drawers are contained, the same tools as in Nº. 6. In another drawer is contained, a set of garden tools, as follows:—a rake, a hoe, an hook bil, a paddle, a hoe, and a fruit knife with hook; these 6 articles made to screw into a feril to fix upon a staff; also a pruning knife, a pair of scissars, a pair of shears, a fork, and some lift.

N. B. *Any of the other chests may be had with a drawer containing a set of garden tools, as in Nº. 7, for  L.   s.   d. more.*

Nº. 13, contains a set of garden tools, as follow:—

A rake, a saw, an hook bill, a paddle, a hoe, and a fruit knife with hook; these 6 articles made to screw into a feril to fix upon a staff; also a hammer, a fork, a pair of scissars, and a partition with some nails and lift.

Nº. 14, contains a set of garden tools, as under:—

A rake, a saw, an hook bill, a paddle, a hoe, a fruit knife, with hook, these 6 articles made to screw into a ferret upon a staff; also a hammer, a fork, a pruning knife, a pair of scissars, a pair of shears, a small hand saw, and a hatchet; 3 gimblets, and a partition with some nails and lift.

☞ *Any of the above chests may be had, mahogany instead of oak, at an extra Price.*

Nº. 15, contains a set of tools in a japan'd case, 6 inches long, as follows:—

A hand pad, hammer, 2 saws, 2 gimblets, 2 awls, 2 turn screws, 1 timer, 1 counter sink, 1 chissel, and 1 gouge, all to fix in the pad.

Nº. 16, contains a set of tools in a japan'd case, 8 inches long, as follows:—

A hand pad, hammer, 2 saws, 2 gimblets, 2 awls, 3 turn screws, 2 chissels, 2 gouges, 1 timer, 1 country sink, half round and flat file, half round and flat rasp, and 1 square file, all to fix in the pad.

Nº. 17, contains the same tools as in Nº. 16, with the addition of a table vice.

Nº. 18, contains the same tools as Nº. 17, only mahogany case.

Nº. 19, contains the same tools as Nº. 16, only mahogany case.

Nº. 20, contains the same tools as Nº. 17, only mahogany case.

Figure 11.—EARLY 19TH CENTURY: THE ADVERTISEMENTS OF TOOLMAKERS indicated the diversity of production. The Castle Hill Works at Sheffield offered to gentlemen 20 choices of tool chests designed to appeal to a wide variety of users and purses. The chest was available in either oak or mahogany, depending on the gentleman's tastes (fig. 49). (Book 87, Cutler and Company, Castle Hill Works, Sheffield. *Courtesy of the Victoria and Albert Museum.*)

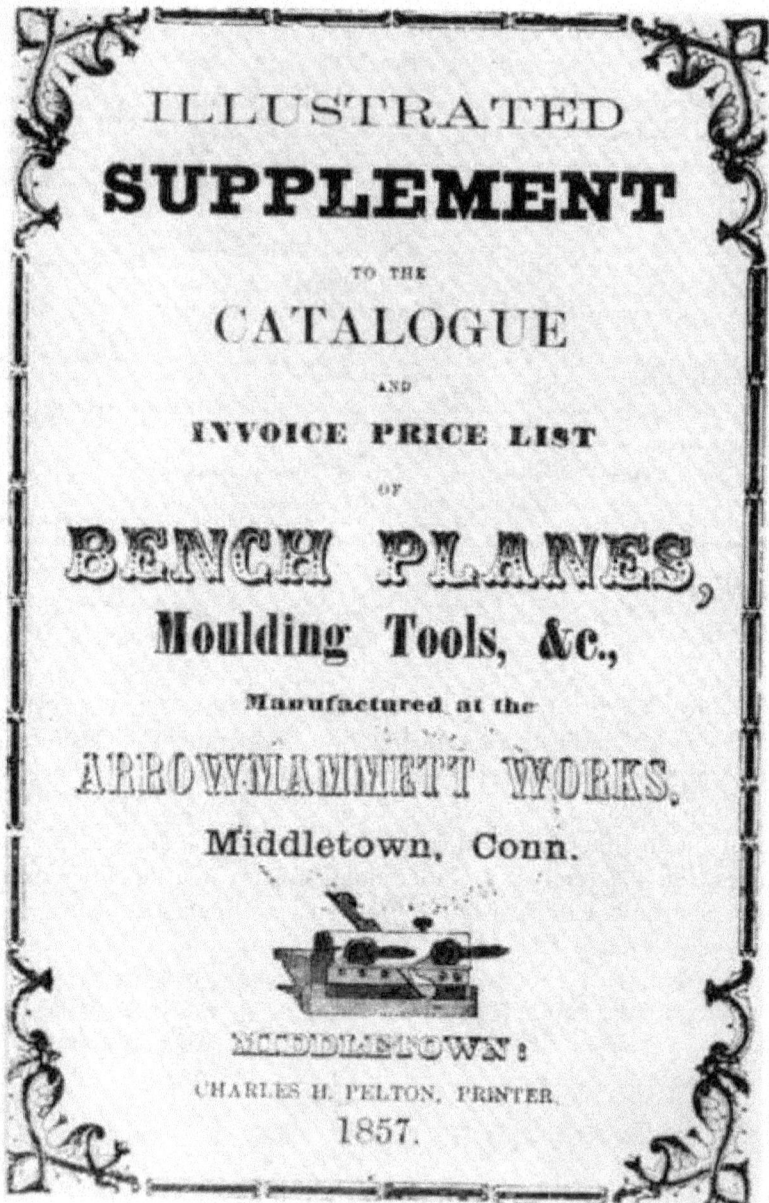

Figure 12.—1857: THE DIVERSITY OF TOOLS available to buyers made necessary the illustrated trade catalogue. Although few in number in the United States before 1850, tool catalogues became voluminous in the last half of the century as printing costs dropped. (Smithsonian Institution Library. Smithsonian photo 49790.)

Peter C. Welsh

American inventories reflect the great increase suggested by the early technical writers and trade catalogues cited above. Compare the content of two American carpenters' shops—one of 1709, in York County, Virginia, and the other of 1827, in Middleborough, Massachusetts. John Crost, a Virginian, owned, in addition to sundry shoemaking and agricultural implements, a dozen gimlets, chalklines, bung augers, a dozen turning tools and mortising chisels, several dozen planes (ogees, hollows and rounds, and plows), several augers, a pair of 2-foot rules, a spoke shave, lathing hammers, a lock saw, three files, compasses, paring chisels, a jointer's hammer, three handsaws, filling axes, a broad axe, and two adzes. Nearly 120 years later Amasa Thompson listed his tools and their value. Thompson's list is a splendid comparison of the tools needed in actual practice, as opposed to the tools suggested by Nicholson in his treatise on carpentry or those shown in the catalogues of the toolmakers.[1] Thompson listed the following:

| | | |
|---|---|---|
| 1 | set bench planes | $6.00 |
| 1 | Broad Axe | 3.00 |
| 1 | Adze | 2.25 |
| 1 | Panel saw | 1.50 |
| 1 | Panel saw | 1.58 |
| 1 | fine do— | 1.58 |
| 1 | Drawing knife | .46 |
| 1 | Trying square | .93 |
| 1 | Shingling hatchet | .50 |
| 1 | Hammer | .50 |
| 1 | Rabbet plane | .83 |
| 1 | Halving do | .50 |
| 1 | Backed fine saw | 1.25 |
| 1 | Inch augre | .50 |
| 1 | pr. dividers or compasses— | .71 |
| 1 | Panel saw for splitting | 2.75 |
| 1 | Tennon gauge | 1.42 |

---

[1] York County Records, Virginia Deeds, Orders, and Wills, no. 13 (1706–1710), p. 248; and the inventory of Amasa Thompson in Lawrence B. Romaine, "A Yankee Carpenter and His Tools," The Chronicle of the Early American Industries Association (July 1953), vol. 6, no. 3, pp. 33–34.

| 1 | Bevel | .84 |
| 1 | Bradd Hammer | .50 |
| 1 | *Architect Book* | 6.50 |
| 1 | Case Mathematical Instruments | 3.62-½ |
| 1 | Panel saw | 2.75 |
| 1 | Grafting saw | 1.00 |
| 1 | Bench screw | 1.00 |
| 1 | Stamp | 2.50 |
| 1 | Double joint rule | .62-½ |
| 1 | Sash saw | 1.12-½ |
| 1 | Oil Can | .17 |
| 1 | Brace & 36 straw cold bits | 9.00 |
| 1 | Window Frame tool | 4.00 |
| 1 | Blind tool | 1.33 |
| 1 | Glue Kettle | .62-½ |
| 1 | Grindstone without crank | 1.75 |
| 1 | Machine for whetting saws | .75 |
| 1 | Tennoning machine | 4.50 |
| | Drafting board and square Bevel— | 1.25 |
| 1 | Noseing sash plane with templets & copes | 4.50 |
| 1 | pr. clamps for clamping doors | 2.17 |
| 1 | Set Bench Planes—double irons.— | 7.50 |
| 1 | Grindstone 300 lbs @ | 6.25 |
| 1 | Stove for shop—$7.25, one elbow .37 & 40 | |
| | lbs second hand pipe $4.00 | 11.62 |
| 1 | Bed moulding | 2.00 |
| 1 | Pr. shears for cutting tin.— | .17 |
| 1 | Morticing Machine | 10.75 |
| 1 | Grecian Ovilo | 1.13 |
| 1-³⁄₁₆ | beed | .67 |
| 1 | Spirit level | 2.25 |
| 1 | Oil stone | .42 |
| 1 | Small trying square | .48 |
| 1 | pareing chisel | .37 |
| 1 | Screw driver | .29 |

| | | |
|---|---|---|
| 1 | Bench screw | .75 |
| 1 | Box rule | .50 |
| 1-$\frac{3}{4}$ | Augre | .41 |
| 11 | Gouges | 1.19 |
| 13 | Chisels | 1.17 |
| 1 | small iron vice | .52 |
| 1 | pr. Hollow Rounds | .86 |
| 4 | Framing chisels | 1.05 |
| 1 | Grove plough & Irons—Sold at 4.50 | 5.00 |
| 1 | Sash plane for 1-$\frac{1}{4}$ stuff | 1.50 |
| 1 | Copeing plane | .67 |
| 1 | Bead $\frac{1}{4}$— | .75 |
| 1 | Bead $\frac{3}{4}$ | 1.00 |
| 1 | Rabbet (Sold at .92) | .92 |
| 1 | Smooth plane | 1.50 |
| 1 | Strike Block | .92 |
| 1 | Compass saw | .42 |
| 6 | Gauges | 1.83 |
| 1 | Dust brush | .25 |
| 1 | Rasp, or wood file | .25 |
| 1 | Augre 2 in. | .76 |
| 1 | Augre 1 in. | .40 |
| 1 | Do $\frac{3}{4}$ | .30 |
| 1 | Spoke shave | .50 |
| 1 | Bevel— | .25 |
| 1 | Box rule | .84 |
| 1 | Iron square | 1.42 |
| 1 | Box rule | 1.25 |
| 1 | Spur Rabbet (Sold—1.17) | 1.33 |
| 1 | Pannel plane | 1.25 |
| 1 | Sash plane | 1.25 |
| 1 | pr. Match planes | 2.25 |
| 1 | Two inch chisel or firmer— | .42 |
| 1 | Morticing chisel $\frac{3}{8}$ | .25 |
| 1 | Large screw driver | 1.00 |
| 1 | Pr. small clamps | .50 |
| 1 | pr. Spring dividers | .92 |
| 1 | do-nippers | .20 |
| 1 | Morticing chisel $\frac{1}{2}$ in. | .28 |
| 1 | Ovilo & Ostrigal $\frac{3}{4}$— | 1.25 |

| | | |
|---|---|---|
| 1 | Scotia & Ostrigal $\frac{5}{8}$— | 1.08 |
| 1 | Noseing— | 1.08 |
| 1 | Pr. Hollow & rounds | 1.33 |
| 1 | Ogee— $\frac{1}{2}$ inch | 1.00 |
| 1 | Ostrigal $\frac{7}{8}$ inch | 1.00 |
| 1 | Bit— | .15 |
| 1 | Beed $\frac{1}{2}$ inch | .83 |
| 1 | Claw hammer | .67 |
| 1 | Fillister | 2.50 |
| 2 | Beeds at $\frac{5}{8}$ | 1.83 |
| 1 | Pair Quirk tools | 1.50 |
| 1 | Side Rabbet plane | .83 |
| 1 | Large steel tongued sq. | 1.71 |
| 1 | Saw & Pad | .67 |
| 1 | pr. fire stones | .50 |
| 1 | small trying sq. | .50 |
| 1 | Set Bench planes double ironed without smooth plane | 6.00 |
| 1 | Bench screw | .75 |

Peter C. Welsh

Figure 13.—EARLY 18TH CENTURY: In addition to their special function and importance as survivals documenting an outmoded
technology, the hand tool often combines a gracefulness of line and a
sense of proportion that makes it an object of great decorative appeal.
The dividers of the builder or shipwright illustrated here are of French
origin and may be valued as much for their cultural significance as for
their technical importance. (Smithsonian photo 49792-G.)

# Woodworking Tools 1600-1900

By 1900, the carpenter's tool chest, fully stocked and fit for the finest craftsman, contained 90 or more tools. Specialization is readily apparent; the change in, and achievement of, the ultimate design of a specific tool is not so easily pinpointed. Only by comparing illustrations and surviving examples can such an evolution be appreciated and in the process, whether pondering the metamorphosis of a plane, a brace and bit, or an auger, the various stages of change encountered coincide with the rise of modern industrial society.

Figure 14.—1688: FRONTISPIECE FROM JOHN BROWN, *The Description and Use of the Carpenter's Rule*, London, 1688. (Library of Congress.)

# Configuration

Hand tools are often neglected in the search for the pleasing objects of the past. Considered too utilitarian, their decorative appeal—the mellow patina of the wood plane or the delicately tapered legs of a pair of dividers—often goes unnoticed. Surprisingly modern in design, the ancient carpenter's or cabinetmaker's tool has a vitality of line that can, without reference to technical significance, make it an object of considerable grace and beauty. The hand tool is frequently a lively and decorative symbol of a society at a given time—a symbol, which, according to the judges at London's Crystal Palace Exhibition in 1851, gives "indications of the peculiar condition and habits of the people whence they come, of their social and industrial wants and aims, as well as their natural or acquired advantages."[1] The hand tool, therefore, should be considered both as an object of appealing shape and a document illustrative of society and its progress.

---

[1] Reports by the Juries: Exhibition of the Works of Industry of All Nations, 1851 (London, 1852), p. 485.

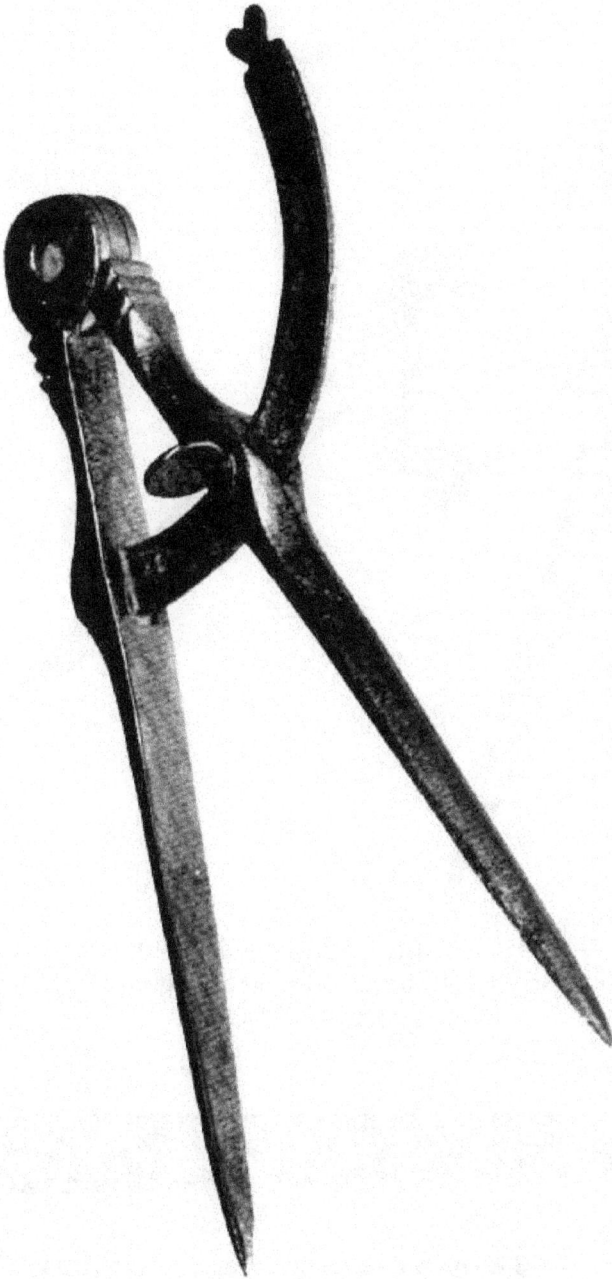

Figure 15.—18TH CENTURY: Cabinetmaker's dividers of English origin. (Private collection. Smithsonian photo 49789-B.)

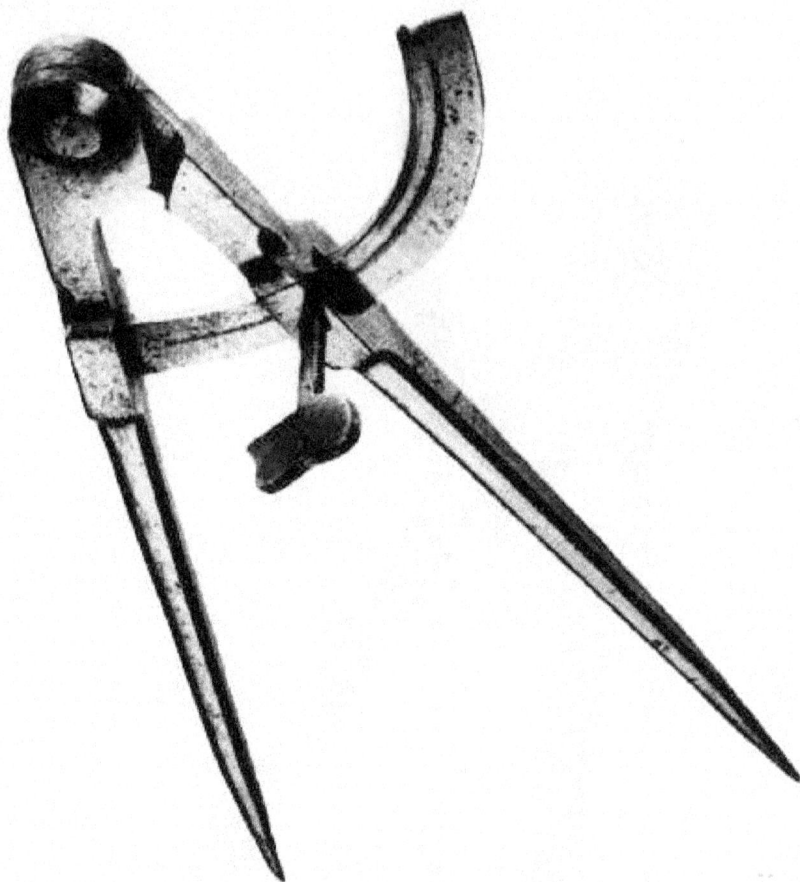

Figure 16.—1783: CABINETMAKER'S dividers of English manufacture, dated, and marked T. Pearmain. See detail, figure 17. (Smithsonian photo 49792-C.)

Figure 17.—1783: DETAIL OF CABINETMAKER'S DIVIDERS showing name and date.

Figure 18.—18TH CENTURY: Carpenter's dividers of English origin, undated. (Smithsonian photo 49792-B.)

On first sight, it is the conformation rather than any facet of its technical or social significance that strikes the eye; perhaps the most decorative of tools are early dividers and calipers which, prior to their standardization, existed in seemingly endless variety. The great dividers used by the shipbuilder and architect for scribing and measuring timbers not only indicate building techniques (accession 61.548) but also document 17th-and early 18th-century decorative metalwork, as seen in figure 13. Well before the 17th century, artists and engravers recognized them as intriguing shapes to include in any potpourri of instruments, either in cartouches or the frontispieces of books (fig. 14).

Figure 19.—1855: THE FRONTISPIECE FROM EDWARD SHAW, *The Modern Architect* (Boston, 1855), shows the carpenter's dividers in the foreground unchanged in form from those illustrated in figure 18. Of further interest in Shaw's plate is the dress of the workmen and the balloon frame of the house under construction. (Smithsonian photo 49792-A.)

Peter C. Welsh

The two pairs of cabinetmaker's dividers illustrated in figures 15 and 16 suggest significant changes in the design of a basic tool. The dividers shown in figure 15 are English and would seem to be of early 18th-century origin, perhaps even earlier. They are Renaissance in feeling with decorated legs and a heart-shaped stop on the end of the slide-arm. In character, they are like the great dividers shown in figure 13: functional, but at the same time preserving in their decoration the features common to a wide variety of ironwork and wares beyond the realm of tools alone. The dividers pictured in figure 16 are a decided contrast. Dated 1783, they are strongly suggestive of Sheffield origin. Gone is the superfluous decoration; in its place is the strong, crisp line of a tool that has reached nearly the ultimate of function and manufacture, a device which both in general appearance and precise design is very modern in execution. Equally intriguing are the smaller, more slender dividers (accession 319557) of the 18th-century house-builder as seen in figure 18, a form that changed very little, if at all, until after 1850—a fact confirmed by the frontispiece of Edward Shaw's *The Modern Architect*, published in Boston in 1855 (fig. 19). The double calipers of the woodturner (fig. 20) have by far the most appealing and ingenious design of all such devices. Designed for convenience, few tools illustrate better the aesthetic of the purely functional than this pair of 19th-century American calipers.

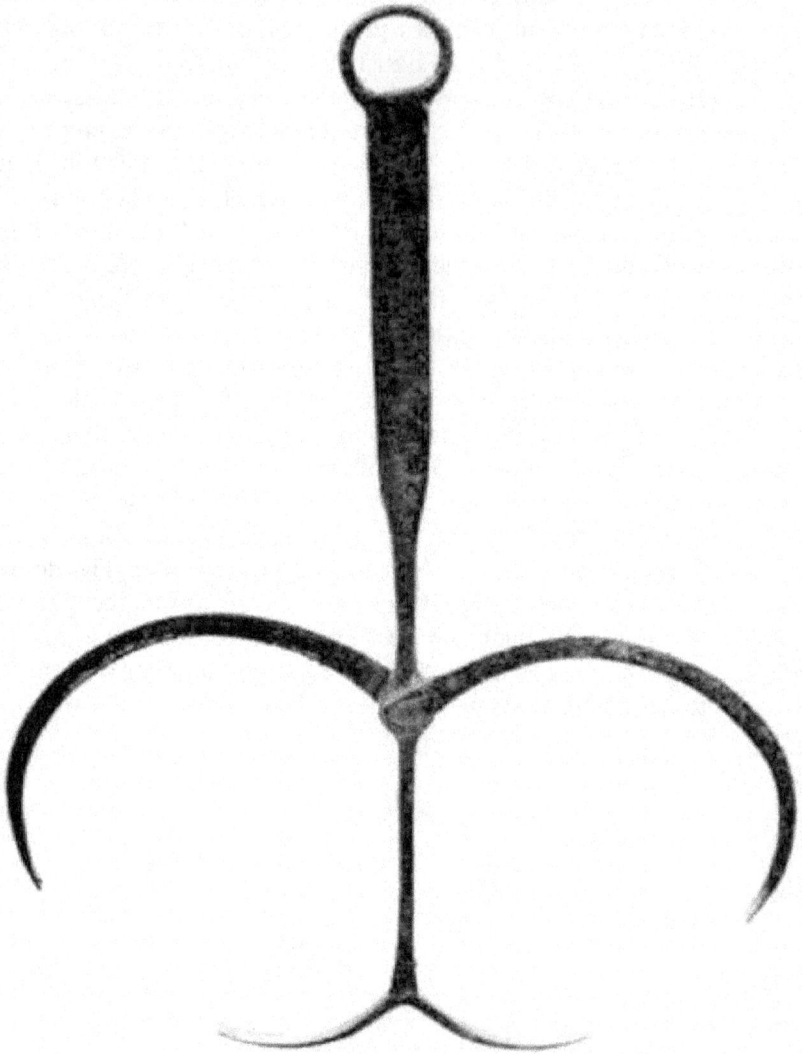

Figure 20.—EARLY 19TH CENTURY: THE DOUBLE CALIPERS of the woodturner permitted double readings to be taken without changing the set of the tool. Inherent in this practical design is a gracefulness of line seldom surpassed. (Private collection. Smithsonian photo 49793-C.)

Peter C. Welsh

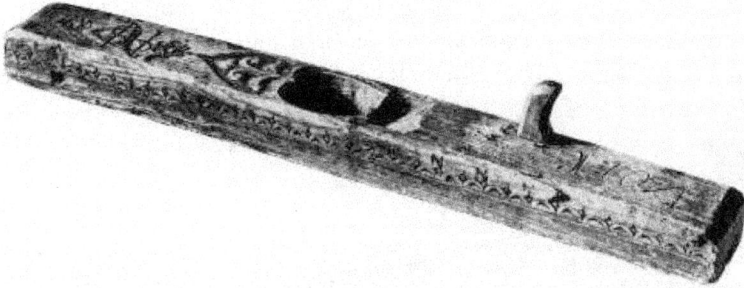

Figure 21.—1704: THE FLOOR PLANE OR LONG JOINER of Norwegian origin exhibits the characteristic decoration of the stock and mouth, patterns common on tools of northern European and Scandinavian origin. (*Courtesy of the Norsk Folkemuseum, Oslo, Norway.*)

Intended to establish proportion and to insure precision, it seems a natural consequence that dividers and calipers should in themselves reflect the same sense of balance and grace that they were designed to govern. Still, even the most prosaic examples of woodworking tools, completely divorced from the quasi-mathematical devices of measure and proportion, have this quality and can be admired solely as decorative objects. This is most evident in the three European bench planes illustrated in figures 21, 22, and 23: one Norwegian, dated 1704; one Dutch (accession 319562), dated 1756; and one German, dated 1809. The Norwegian and German examples, with their elaborately carved bodies and heart-shaped mouths, are typical of the type that Swedish and German colonists in America might have used in the 17th and 18th centuries. They are important for that reason. Also, all three exhibit elaboration found on other material survivals from these countries in their respective periods. For example, the incised rosette of the Dutch plane (fig. 22) is especially suggestive of the rosettes found on English and American furniture of the 1750's and 1760's, specifically on high chests.

The decorative motifs that characterized European tools of the 17th and 18th centuries obscured technical improvement. By contrast, in England and America, tools gained distinction through the directness of their design. Following English patterns, tools of American make were straightforward. Only later, in new tool types, did they imitate the rococo flourish of their European predecessors. In America, as in England, the baroque for things functional seemingly had little appeal. This is particularly true of woodworking planes on which, unlike their continental cousins, embellishment is rarely seen. Exemplifying this

33

tradition are three early 19th-century American planes: a plow, for cutting channels of various widths on board edges, marked "G. White, Phild$^a$" (fig. 24); a rabbet, for notching the margin of boards; made by E.W. Carpenter of Lancaster, Pennsylvania (fig. 25); and a jack or foreplane, for rough surfacing (accession 61.547), made by A. Klock and dated 1818 as seen in figure 26.

Figure 22.—1756: THE HIGHLY elaborated stock and rosette-incised wedge of the smoothing plane recall the decoration on furniture of the period. The plane is of Dutch origin. (Smithsonian photo 49792-F.)

Figure 23.—1809: THIS BENCH PLANE of German origin is dated 1809. It is of a traditional form that persists to the present day. The planes pictured in figures 21, 22, and 23 are similar to the type brought to North America by non-English colonists. (Private collection. Smithsonian photo 49793-F.)

Peter C. Welsh

Figure 24.—ABOUT 1818: This plow plane, used to cut narrow channels on the edges of boards, was made by G. White of Philadelphia in
the early 19th century. It is essentially the same tool depicted in the
catalogues of Sheffield manufactures and in the plates from Martin
and Nicholson. The pattern of the basic bench tools used in America
consistently followed British design, at least until the last quarter of
the 19th century. (Private collection. Smithsonian photo 49794-E.)

Figure 25. 1830–1840: THE DESIGN of the rabbet plane, used to cut a groove of fixed width and depth on the edge of a board, was not improved upon in the 19th century. The carpenter's dependence on this tool lessened only after the perfection of multipurpose metallic planes that could be readily converted to cut a "rabbet." (Private collection. Smithsonian photo 494789-H).

The question of dating arises, since only the Klock piece is firmly fixed. How, for example, is the early 19th-century attribution arrived at for the planes inscribed White and Carpenter? First, the nature of the stamped name "G. White" is of proper character for the period. Second, G. White is listed in the Philadelphia city directories as a "plane-maker" between the years 1818 and 1820, working at the back of 5 Filbert Street and later at 34 Juliana Street. Third, internal evidence on the plane itself gives a clue. In this case, the hardware—rivets and furrels—is similar if not identical to that found on firearms of the period, weapons whose dates of manufacture are known. The decorative molding on the fence of this plane is proper for the period; this is not a reliable guide, however, since similar moldings are retained throughout the century. Finally, the plane is equipped with a fence controlled by slide-arms, fixed with wedges and not by adjustable screw arms. After 1830, tools of high quality, such as White's, invariably have the screw arms. The rabbet plane, made by Carpenter, is traceable via another route, the U.S. Patent Office records. Carpenter, self-designated "toolmaker of Lancaster," submitted patents for the improvement of wood planes between 1831 and 1849. Examples of Carpenter's work, always stamped as shown in figure 27, survive, both dated and undated. There are several of his planes in the collections of the Bucks County Historical Society, and dated pieces are known in private collections.

Inherent in the bench planes is a feeling of motion, particularly in the plow and the rabbet where basic design alone conveys the idea that they were meant to move over fixed surfaces. Of the three examples, only the brass tippings and setscrew of the plow plane suggest any enrichment, and of course these were not intended for decoration; in later years, however, boxwood, fruitwood, and even ivory tips were added to the more expensive factory models. Also unintentional, but pleasing, is the distinctive throat of the rabbet plane—a design that developed to permit easy discharge of shavings, and one that mass manufacture did not destroy.

Figure 26.—1818: THE JACK PLANE, used first by the carpenter for rapid surfacing, is distinguished primarily by the bezeled and slightly convex edge of its cutting iron. As with the plow and the rabbet, its shape is ubiquitous. Dated and marked A. Klock, this American example follows precisely those detailed in Sheffield pattern books. (Smithsonian photo 49794-C.)

Figure 27.—1830–1840: DETAIL OF the rabbet plane (fig. 25) showing the characteristic stamp of E.W. Carpenter. (Smithsonian photo 49794-D.)

Figure 28.—ABOUT 1631: THE PRECEDING ILLUSTRATIONS emphasize
the divergent appearance of European and Anglo-American tools.
This, however, was not always the case. The woodworker's shop by
the Dutch engraver Jan Van Vliet suggests the similarity between English and European tool types in the 17th century. Note in particular the
planes, axe, brace, and auger as compared to Moxon. (Library of Congress, Division of Prints and Photographs.)

Figure 29.—1690: THE CABINETMAKER'S SHOP from Elias Pozelius, *Orbus Pictus nach Zeichnugen der Susanna Maria Sandrart*, Nürnberg, 1690. (Library of Congress.)

Figure 30.—1568: THE WOODWORKER'S SHOP from Hans Sachs, *Eygentliche Beschrerbung Aller Stande ... mit Kunstreichen Figuren* [by Jost Amman], Frankfurt, 1568. (Library of Congress.)

The divergence from European to an Anglo-American hand-tool design and the approximate date that it occurred can be suggested by a comparison of contemporary illustrations. The change in the wooden bench plane can be followed from the early 17th century through its standardization at the end of the 18th century. Examine first the planes as drawn in the 1630's by the Dutchman Jan Van Vliet (fig. 28), an etcher of Rembrandt's school at Leiden, and also the examples illustrated by Porzelius (fig. 29) and by Jost Amman (fig. 30). Compare them to Moxon's plate (fig. 31) from the *Mechanick Exercises* (3rd ed., 1703) and to the splendid drawing of the bench plane from André-Jacob Roubo's *L'Art du menuisier*, published in 1769 (fig. 32). In all of them, the rounded handle, or tote, and the fore-horn appear, characteristics of both European and English planes of the period before 1750. The similarity ends with the mass production of hand tools from the shops of the English toolmaking centers, principally Sheffield. An illustration from a pattern and design book of the Castle Hill Works, Sheffield, dating from the last quarter of the 18th century (fig. 33), shows the achieved, familiar form of the bench planes, as well as other tools. The use of this form in America is readily documented in Lewis Miller's self-portrait while working at his trade in York, Pennsylvania, in 1810 (fig. 34) and by the shop sign carved by Isaac Fowle in 1820 for John Bradford (fig. 35). In each example, the bench plane clearly follows the English prototype.

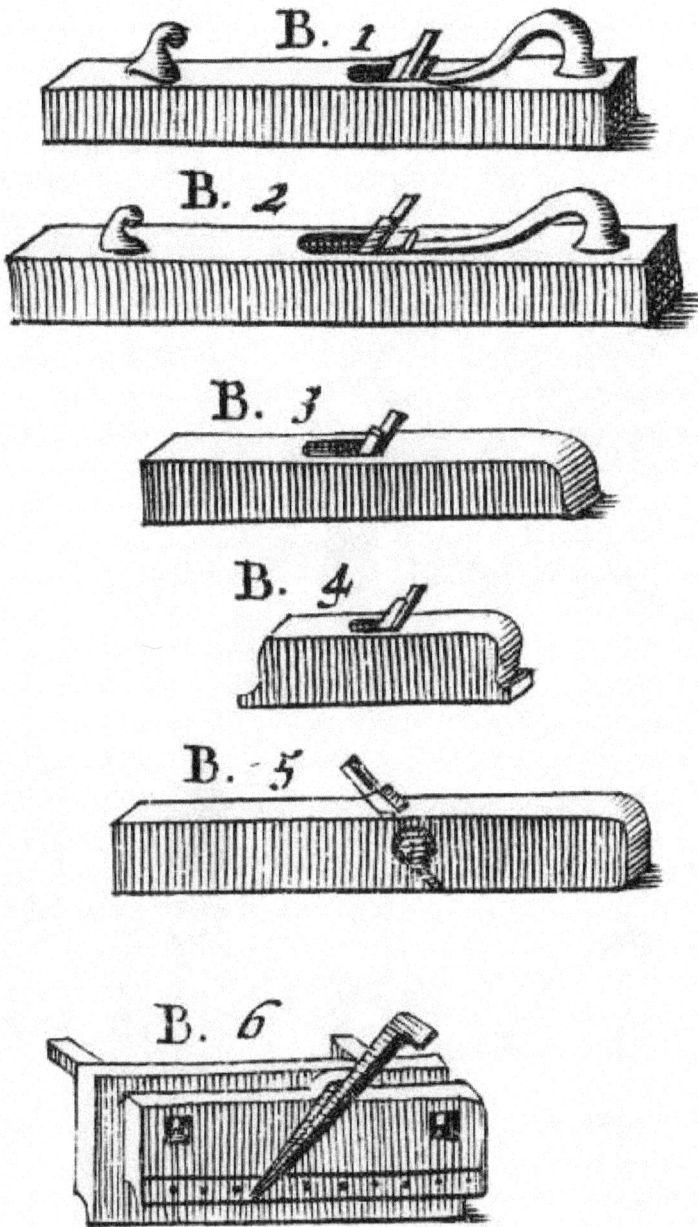

Figure 31.—1703: DETAIL OF THE BENCH PLANES from Moxon's
*Mechanick Exercises*.

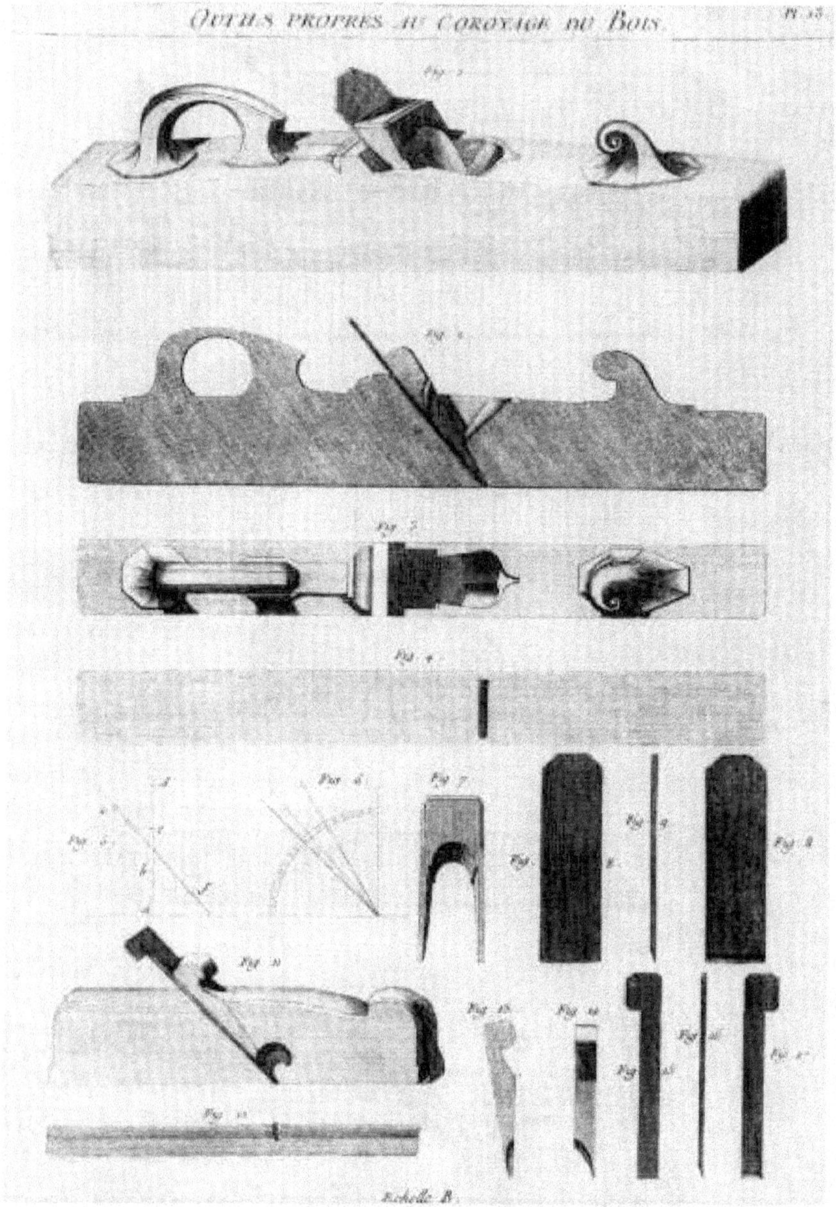

Figure 32.—1769: ANDRÉ-JACOB ROUBO'S PRECISE RENDERING of the
bench plane retains the essential features shown by Moxon—the
rounded tote or handle and the curved fore-horn. (André-Jacob Roubo,
*L'Art du menuisier*, 1769.)

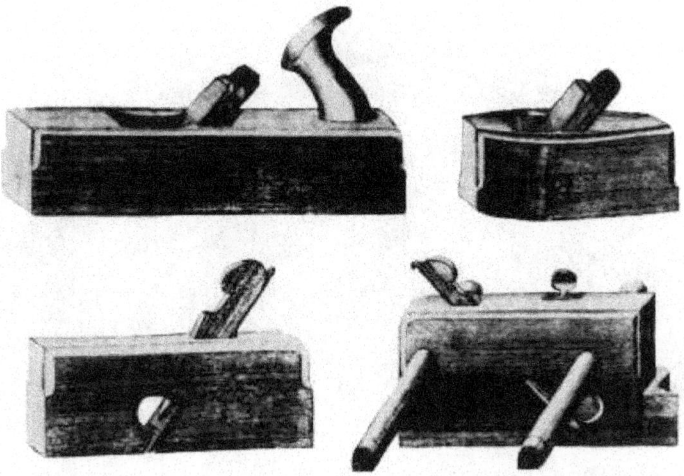

Figure 33.—EARLY 19TH CENTURY: The bench plane illustrated in Roubo or Moxon is seldom seen in American tool collections. The bench planes, smoothing planes, rabbets, and plows universally resemble those shown in this illustration from the pattern book of the Castle Hill Works, Sheffield. (Book 87, Cutler and Company, Castle Hill Works, Sheffield. *Courtesy of the Victoria and Albert Museum.*)

Figure 34.—ABOUT 1810: LEWIS MILLER WORKING AT HIS BENCH in York, Pa. In a predominantly Pennsylvania-German settlement, the plane used by Miller conforms to the Sheffield type illustrated in the catalogue of the Castle Hill Works as shown in figure 33. (York County Historical Society, York, Pa.)

Figure 35.—1820: JOHN BRADFORD'S shop sign carved by Isaac Fowle is a unique documentary of early 19th-century tool shapes and is in the Bostonian Society, Boston, Mass. (Index of American Design, The National Gallery, Washington, D.C.)

Peter C. Welsh

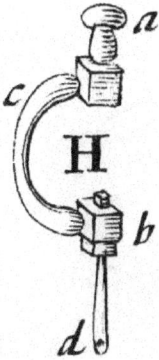

Figure 36.—1703: THE JOINER'S brace and bit—a detail from Moxon, *Mechanick Exercises* ..., London, 1703. (Library of Congress, Smithsonian photo 56635.)

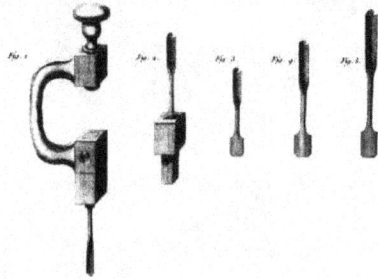

Figure 37.—1769: ROUBO'S ILLUSTRATION OF THE BRACE and bit differs from Moxon's only in the precision of the delineation. Contrast this form with that of the standard Sheffield version in figure 38 and the metallic braces illustrated in figures 40 through 44. From these plates can be seen the progression of the bitstock toward its ultimate perfection in the late 19th century. (André-Jacob Roubo, *L'Art du menuisier*, 1769.)

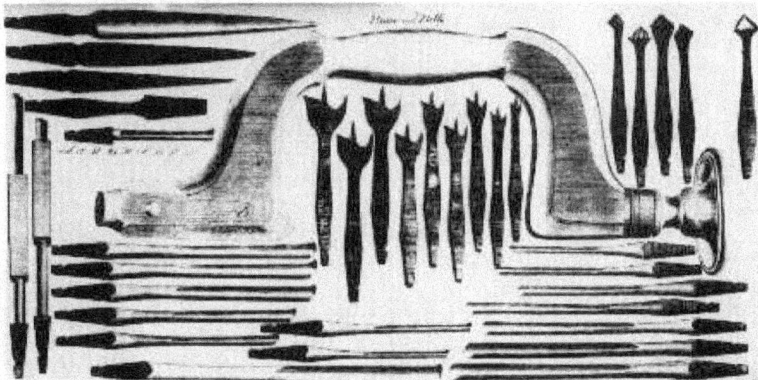

Figure 38.—EARLY 19TH CENTURY: THE MASS-PRODUCED VERSION of the wooden brace and bit took the form illustrated in Book 87 of Cutler's Castle Hill Works. (*Courtesy of the Victoria and Albert Museum.*)

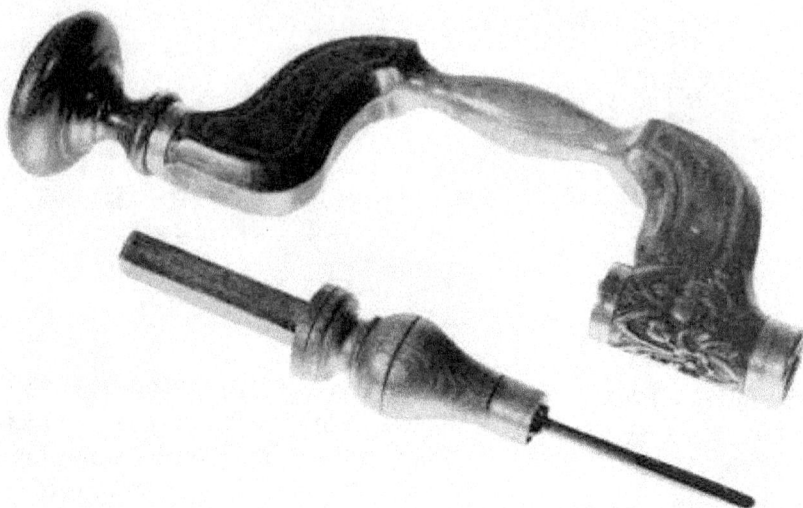

Figure 39.—18TH CENTURY: THE TRANSITIONAL FORM of the wooden brace and bit incorporated the overall shape of the mass-produced version but retained the archaic method of fastening the bit to the chuck. The tool is of Dutch origin and suggests the influence of Sheffield design on European tools. (Smithsonian photo 49792-E.)

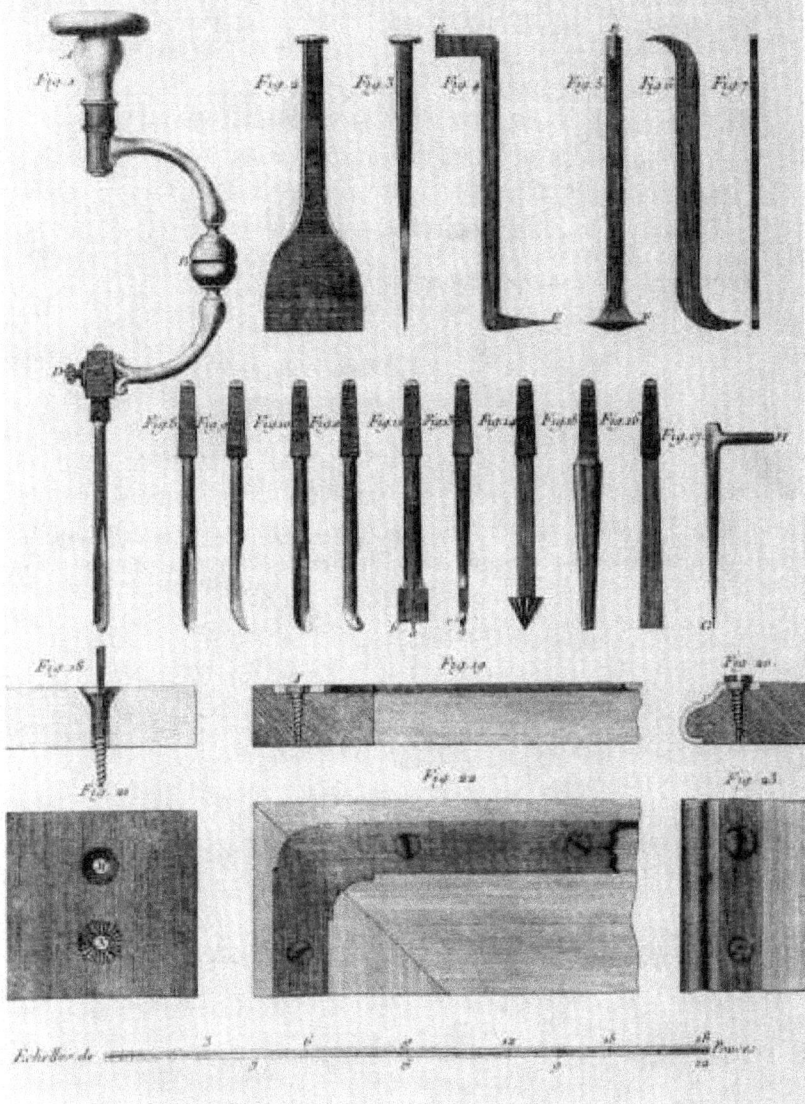

Figure 40.—1769: ROUBO ILLUSTRATED THE METALLIC BRACE and, in addition, suggested its use as a screwdriver. (André-Jacob Roubo, *L'Art du menuisier*,1769.)

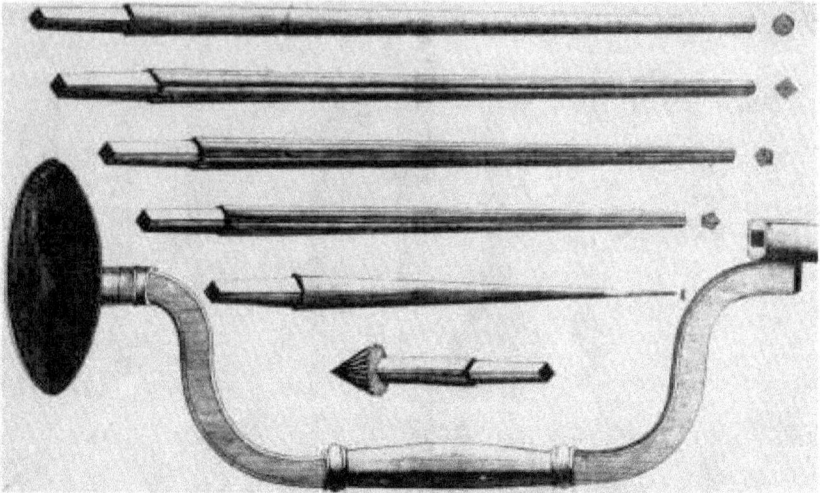

Figure 41.—ABOUT 1775: FORD, WHITMORE AND BRUNTON made and sold clockmaker's braces of metal with a sweep and shank that was imitated by American patentees in the 19th century. (Catalogue of Ford, Whitmore and Brunton, Birmingham, England. *Courtesy of the Birmingham Reference Library.*)

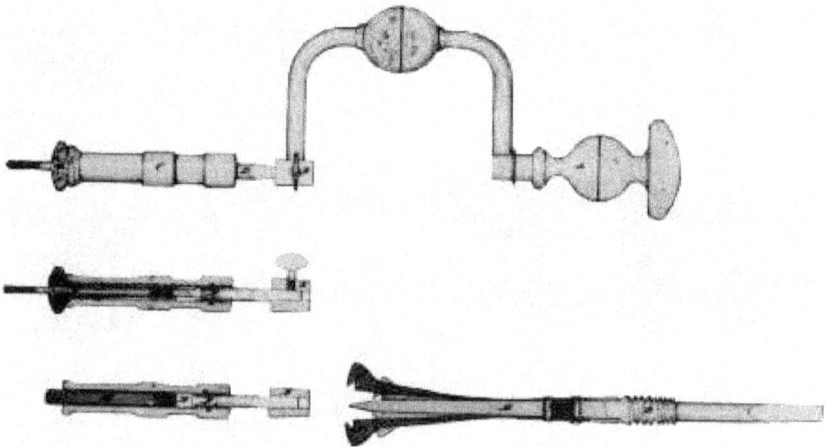

Figure 42.—1852: NEARLY ONE HUNDRED YEARS after Roubo's plate appeared, Jacob Switzer applied for a patent for an "Improved Self Holding Screw Driver." The similarity of Switzer's drawing and Roubo's plate is striking. (Original patent drawing 9,457, U.S. Patent Office, Record Group 241, the National Archives.)

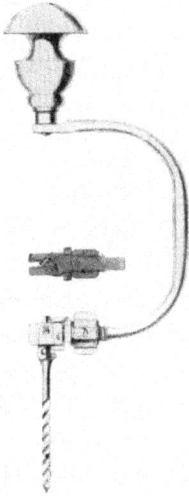

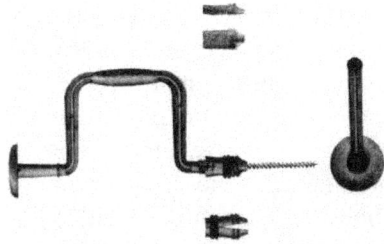

Figure 44.—1865: MILTON NOBLES' PATENT perfecting the chuck which held the auger bit was an important step along the path which led ultimately to the complete acceptance of the metallic brace. Barber's ratchet brace shown in figure 66 completes the metamorphosis of this tool form in the United States. (Original patent drawing 51,660, U.S. Patent Office, Record Group 241, the National Archives.)

Figure 43.—1866: THE SIMPLICITY AND STRENGTH of the brace proposed by J. Parker Gordon is in sharp contrast to the heavily splinted sides of the wooden brace commonly used in mid-19th-century America. (Original patent drawing 52,042, U.S. Patent Office, Record Group 241, the National Archives.)

The carpenter's brace is another instance of divergent design after a common origin. Refer again to Van Vliet's etching of the woodworker's shop (fig. 28), to the detail from Moxon (fig. 36), and from Roubo (fig. 37). All show the brace in a form familiar since the Middle Ages, a shape common to both delineators and craftsmen of the Continent and the British Isles. But, as the plane changed, so changed the brace. The standard form of this tool as it was used and produced in the United States in the 19th century can be seen in another plate from the catalogue of the Castle Hill Works at Sheffield (fig. 38). This English influence on American tool design is no surprise, since as early as 1634 William Wood in *New England's Prospect* suggested that colonists take to the New World "All manner of Ironwares, as all manner of nailes for houses ... with Axes both broad and pitching ... All man-

ners of Augers, piercing bits, Whip-saws, Two handed saws, Froes ...,
rings for Bettle heads, and Iron-wedges."

English tool design in the 18th century also influenced the continental toolmakers. This can be seen in figure 39 in a transitional-type
bitstock (accession 319556) from the Low Countries. Adopting an
English shape, but still preserving the ancient lever device for holding
the bit in place, the piece with its grapevine embellishment is a marked
contrast to the severely functional brass chucks on braces of English
manufacture. No less a contrast are metallic versions of the brace.
These begin to appear with some regularity in the U.S. patent specifications of the 1840's; their design is apparently derived from 18th-
century precedents. Roubo (fig. 40) illustrated a metal bitstock in
1769, as did Ford, Whitmore & Brunton, makers of jewelers' and
watchmakers' tools, of Birmingham, England, in their trade catalogue
of 1775 (fig. 41). Each suggests a prototype of the patented forms of
the 1840's. For example, in 1852, Jacob Switzer of Basil, Ohio, suggested, as had Roubo a hundred years earlier, that the bitstock be used
as a screwdriver (fig. 42); but far more interesting than Switzer's idea
was his delineation of the brace itself, which he described as "an ordi-
nary brace and bit stock" (U.S. pat. 9,457). The inference is that such a
tool form was already a familiar one among the woodworking trades in
the United States. Disregarding the screwdriver attachment, which is
not without merit, Switzer's stock represents an accurate rendering of
what was then a well-known form if not as yet a rival of the older
wooden brace. Likewise, J. Parker Gordon's patent 52,042 of 1866
exemplifies the strengthening of a basic tool by the use of iron (fig.
43) and, as a result, the achievement of an even greater functionalism
in design. The complete break with the medieval, however, is seen in a
drawing submitted to the Commissioner of Patents in 1865 (pat.
51,660) by Milton V. Nobles of Rochester, New York.[1] Nobles' crea-
tion was of thoroughly modern design and appearance in which, unlike
earlier types, the bit was held in place by a solid socket, split sleeve,
and a tightening ring (fig. 44). In three centuries, three distinct design
changes occurred in the carpenter's brace. First, about 1750, the so-
called English or Sheffield bitstock appeared. This was followed in the
very early 19th century by the reinforced English type whose sides
were splinted by brass strips. Not only had the medieval form largely

---

[1] U.S. patent specifications cited in this paragraph may be found at the U.S.
Patent Office, Washington, D.C.

disappeared by the end of the 18th century, but so had the ancient lever-wedge method of fastening the bit in the stock, a device replaced by the pressure-spring button on the side of the chuck. Finally, in this evolution, came the metallic stock, not widely used in America until after the Civil War, that embodied in its design the influence of mass manufacture and in its several early versions all of the features of the modern brace and bit.

Figure 45.—19TH CENTURY: THE UPHOLSTERER'S HAMMER is an unknown; it is not dated, its maker is anonymous, as is its user. It is of American origin, yet of a style that might have been used in England or on the Continent. This lack of provenance need not detract from its significance as a material survival. This hammer, the brace (fig. 46), the bevel (fig. 47), and the compass saw (fig. 48) are sufficiently provocative in their design to conjure some image of a technology dependent upon the skilled hand of craftsmen working in wood and of the relationship between the hand, the tool, and the finished product. (Smithsonian photo 49793-A.)

Figure 46.—18TH CENTURY: THE BRACE AND BIT in its nonfactory form conforms to a general design pattern in which none of the components are ever precisely alike. This aspect of variety of detail—sophistication, crudeness, decorative qualities or the like—reflects something of the individuality of the toolmaker, a quality completely lost in the standardization of the carpenter's brace. (Smithsonian photo 49794-A.)

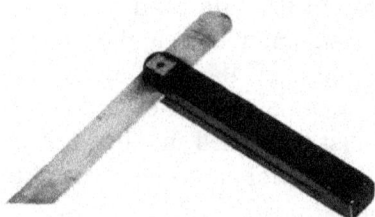

Figure 47.—18TH CENTURY: The visually pleasing qualities of walnut and brass provide a level of response to this joiner's bevel quite apart from its technical significance. (Private collection. Smithsonian photo 49793-B.)

Figure 48.—18TH CENTURY: THE HANDLE OF THE COMPASS SAW, characteristically Dutch in shape, is an outstanding example of a recurring functional design, one which varied according to the hand of the sawer. (Smithsonian photo 49789-C.)

Henry Ward Beecher, impressed by the growing sophistication of the toolmakers, described the hand tool in a most realistic and objective manner as an "extension of a man's hand." The antiquarian, attuned to more subjective and romantic appraisals, will find this hardly sufficient. Look at the upholsterer's hammer (accession 61.35) seen in figure 45: there is no question that it is a response to a demanding task that required an efficient and not too forceful extension of the workman's hand. But there is another response to this implement: namely, the admiration for an unknown toolmaker who combined in an elementary striking tool a hammerhead of well-weighted proportion to be wielded gently through the medium of an extremely delicate handle. In short, here is an object about whose provenance one need know very little in order to enjoy it aesthetically. In a like manner, the 18th-century bitstock of Flemish origin (fig. 46), the English cabinetmaker's bevel of the same century (fig. 47), and the compass saw (accession 61.52, fig. 48) capture in their basic design something beyond the functional extension of the craftsman's hand. The slow curve of the bitstock, never identical from one early example to another, is lost in later factory-made versions; so too, with the coming of cheap steel, does the combination of wood (walnut) and brass used in the cabinetmaker's bevel slowly disappear; and, finally, in the custom-fitted pistol-like grip of the saw, there is an identity, in feeling at least, between craftsman and tool never quite achieved in later mass-produced versions.

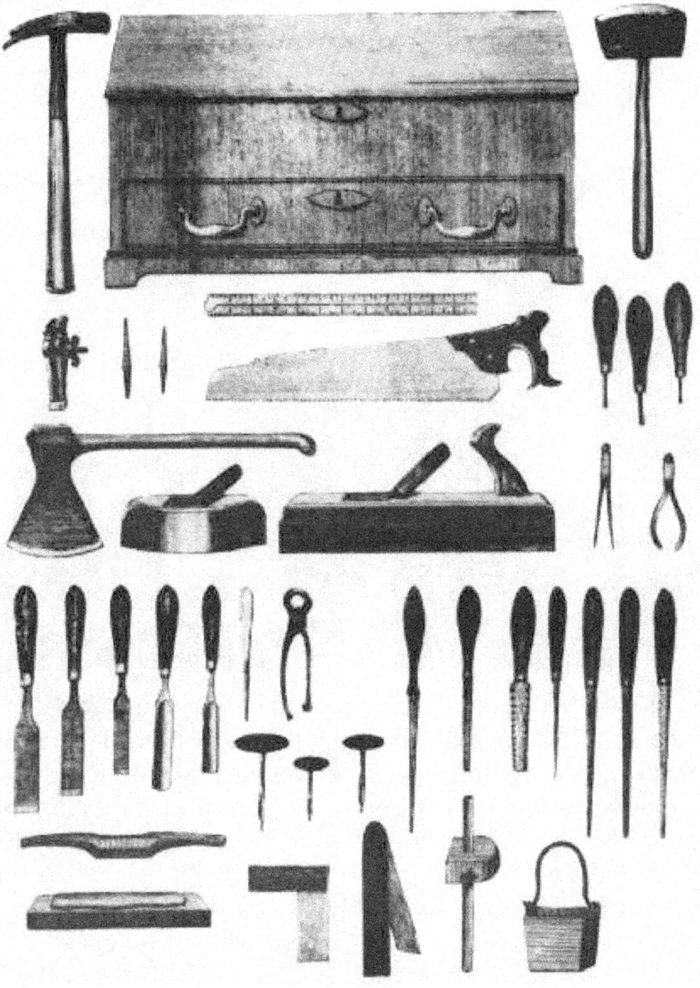

Figure 49.—EARLY 19TH CENTURY: THE DESIGNATION "GENTLEMAN'S TOOL CHEST" required a chest of "high-style" but necessitated no change in the tools it held. (Book 87, Cutler and Company, Castle Hill Works, Sheffield. *Courtesy of the Victoria and Albert Museum.*)

Figure 50.—19TH CENTURY: THE SCREWDRIVER, which began to appear regularly on the woodworker's bench after 1800, did not share the long evolution and tradition of other Anglo-American tool designs. The screwdriver in its early versions frequently had a scalloped blade for no other purpose than decoration. (Smithsonian photo 49794.)

Figure 51.—1870: THE USE of a new material prompted a departure from the traditional in shape and encouraged surface elaboration. The tendency, however, was short lived and the mass-produced metallic plane rapidly achieved a purity of design as pleasing as its wooden predecessors. (Private collection. Smithsonian photo 49789.)

Occasionally, ruling taste is reflected in the design of the carpenter's equipment. Notable is the "gentleman's tool chest" (fig. 49) advertised in the pattern book of the Castle Hill Works. The bracket feet, brass pulls, and inlaid keyholes imitate the style of the domestic chest of drawers of the period 1790 to 1810—undoubtedly, features included by the manufacturer to appeal to a gentleman of refined taste. In contrast to this Sheffield product is the plate from Shaw's *The Modern Architect*. The concept of the builder-carpenter as a gentleman still prevails, although the idea in this American scene is conveyed in the mid-19th century through fashionable dress. The tools and in particular the tool chest reflect only the severest of functional lines (fig. 19, p. 196).

In deference to ruling taste, some tools lost for a time the clean lines that had long distinguished them. The screwdriver, simple in shape (accession 61.46) but in little demand until the 1840's, occasionally became most elaborate in its factory-made form (fig. 50) and departed noticeably from the unadorned style of traditional English and American tools. The scalloped blade, influenced by the rival styles rather than a technical need, seemed little related to the purpose of the tool.[1] No less archaic in decoration was the iron-bodied version of the plow plane (fig. 51). The Anglo-American tradition seems completely put aside. In its place is a most functional object, but one elaborately covered with a shell and vine motif! Patented in 1870 by Charles Miller and manufactured by the Stanley Rule and Level Company, this tool in its unadorned version is of a type that was much admired by the British experts at Philadelphia's Centennial Exhibition in 1876. What prompted such superfluous decoration on the plow plane? Perhaps it was to appeal to the flood of newly arrived American craftsmen who might find in the rococo something reminiscent of the older tools they had known in Europe. Perhaps it was simply the transference to the tool itself of the decorative work then demanded of the wood craftsmen. Or was it mainly a compulsion to dress, with little effort, a lackluster material that seemed stark and cold to Victorians accustomed to the ornateness being achieved elsewhere with the jigsaw and wood? Whatever the cause, the result did not persist long as a guide to hand-tool design. Instead, the strong, plain lines that had evolved over two centuries won universal endorsement at the Centennial Exhibition. The prize tools reflected little of the ornateness apparent in the wares of most of the other exhibitors. American makers of edge tools exhibiting at the Centennial showed the world not only examples of quality but of attractiveness as well.

---

[1] In 1865 George Parr in his application for an improved screwdriver stated categorically that the scalloped blade served no purpose other than decoration. See U.S. patent 45,854, dated January 10, 1865.

HANDLED AXES.

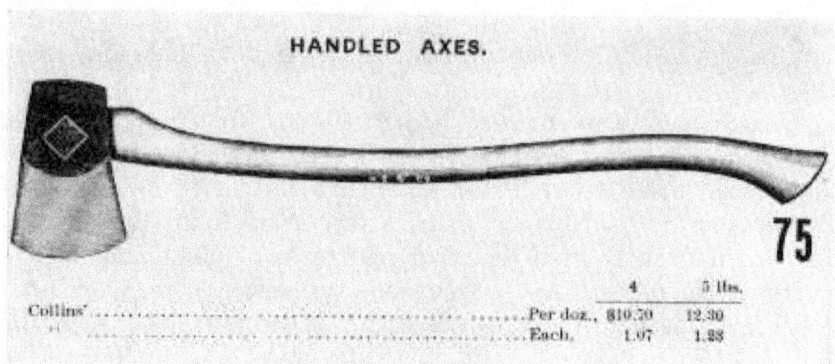

| | 4 | 5 lbs. |
|---|---|---|
| Collins' .......... Per doz., | $10.70 | 12.30 |
| .......... Each. | 1.07 | 1.33 |

Figure 52.—19TH CENTURY: THE AMERICAN AXE WAS UNEXCELLED in design and ease of use. European observers praised it as distinctly American. At the Centennial Exhibition in 1876 Collins and Company of New York City was singled out as one of the outstanding manufacturers exhibiting these axes, a reputation that persisted. (*Tools for all Trades*, Hammacher, Schlemmer and Company, New York, 1896. Smithsonian photo 56625.)

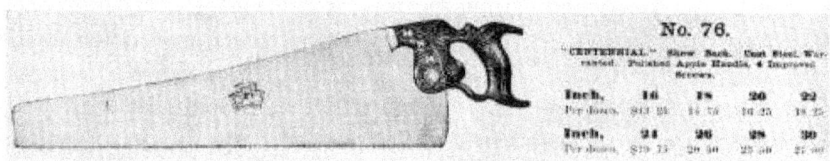

No. 76.

"CENTENNIAL." Skew Back. Cast Steel. Warranted. Polished Apple Handle, 4 Improved Screws.

| Inch, | 16 | 18 | 20 | 22 |
|---|---|---|---|---|
| Per doz., | $13 25 | 14 75 | 16 25 | 18 25 |
| Inch, | 24 | 26 | 28 | 30 |
| Per dozen, | $19 75 | 20 50 | 23 50 | 27 00 |

Figure 53.—1876: DISSTON AND SONS LONG CONTINUED to remind prospective buyers of the company's success at the Philadelphia Centennial Exhibition by retaining the "Centennial Saw, No. 76" as a brand name. (*Illustrated Catalogue*, Baldwin, Robbins and Company, Boston, 1894. Smithsonian photo 56627.)

Peter C. Welsh

# Change

American hand tools in 1876 did not achieve the popular acclaim accorded the Corliss engine, yet few products shown by American exhibitors were more highly praised by foreign experts. It seems justified to suggest that American edge tools displayed at the Centennial had reached their high point of development—a metamorphosis that began with the medieval European tool forms, moved through a period of reliance on English precedents, and ended, in the last quarter of the 19th century, with the production of American hand tools "occupying an enviable position before the world."[1]

Figure 54.—1809: THE INTRODUCTION of the gimlet-pointed auger followed Ezra L'Hommedieu's patent of 1809. From this date until its general disuse in the early 20th century, the conformation of the tool remained unchanged, although the quality of steel and the precision of the twist steadily improved. (Wash drawing from the restored patent drawings awarded July 31, 1809, U.S. Patent Office, Record Group 241, the National Archives. Smithsonian photo 49790-A.)

---

[1] Francis A. Welk, ed., United States Centennial Commission, International Exhibition, 1876, Reports and Awards, Group XV (Philadelphia, 1877), p. 5.

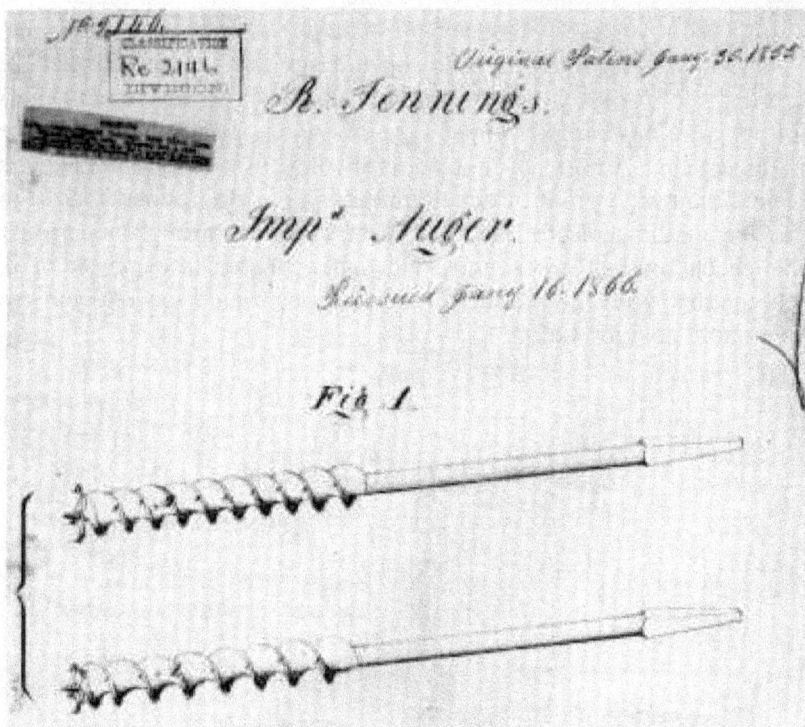

Figure 55.—1855: RUSSELL JENNINGS' improved auger bits, first patented in 1855, received superior citation at the Philadelphia Centennial; in the years following, the trade name "Jennings" was seldom omitted from trade catalogues. (Original wash drawing, patent drawing submitted by R. Jennings, U.S. Patent Office, Record Group 241, the National Archives.)

The tool most highly praised at Philadelphia was the American felling axe (fig. 52) "made out of a solid piece of cast steel" with the eye "punched out of the solid." When compared to other forms, the American axe was "more easily worked," and its shape permitted an easier withdrawal after striking.[1]

Sawmakers, too, were singled out for praise—in particular Disston & Sons (fig. 53) for "improvements in the form of the handles, and in the mode of fixing them to the saw." The Disston saw also embodied

---

[1] Ibid., p. 6.

an improved blade shape which made it "lighter and more convenient by giving it a greater taper to the point." Sheffield saws, once supplied to most of the world, were not exhibited at Philadelphia, and the British expert lamented that our "monopoly remains with us no longer."[1]

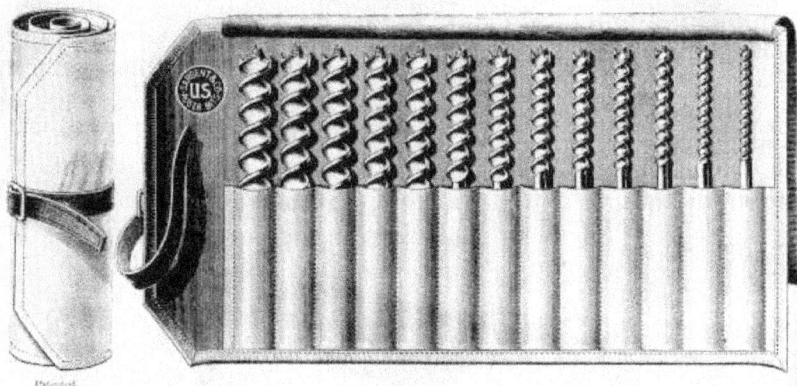

" U. S." Auger Bits in Jennings " Protection " Rolls.

No. 455, Improved Lip and Spur.

No. 488, Jennings Pattern, Extension Lip.

Figure 56.—1894: THE PERSISTENCE OF "JENNINGS" AS A TRADE NAME is suggested by the vignette from the "Illustrated Catalogue" of Baldwin, Robbins and Company, published in 1894. (Smithsonian photo 56628.)

Augers, essential to "the heavier branches of the building trade ... [and] in the workshops of joiners, carpenters, cabinetmakers, turners, carvers, and by amateurs and others," were considered a "most important exhibit" at the Centennial. The auger had attained a perfection in "the accuracy of the twist, the various forms of the cutters, the quality of the steel, and fine finish of the twist and polish." The ancient pod or shell auger had nearly disappeared from use, to be replaced by "the screwed form of the tool" considerably refined by comparison to L'Hommedieu's prototype, patented in 1809 (fig. 54). Russell Jennings' patented auger bits (figs. 55–56) were cited for their "workmanship and quality," and, collectively, the Exhibition "fully established the reputation of American augers."[2] Likewise, makers of braces and bits

---

[1] Ibid., pp. 9–10.

[2] Ibid., pp. 11–12.

were commended for the number of excellent examples shown. Some were a departure from the familiar design with "an expansive chuck for the bit," but others were simply elegant examples of the traditional brace, in wood, japanned and heavily reinforced with highly polished brass sidings. An example exhibited by E. Mills and Company, of Philadelphia, received a certification from the judges as being "of the best quality and finish" (fig. 57). The Mills brace, together with other award-winning tools of the company—drawknives, screwdrivers, and spokeshaves—is preserved in the collections of the Smithsonian Institution (accession 319326). Today as a group they confirm "the remarkably fine quality of ... both iron and steel" that characterized the manufacture of American edge tools in the second half of the 19th century.[1]

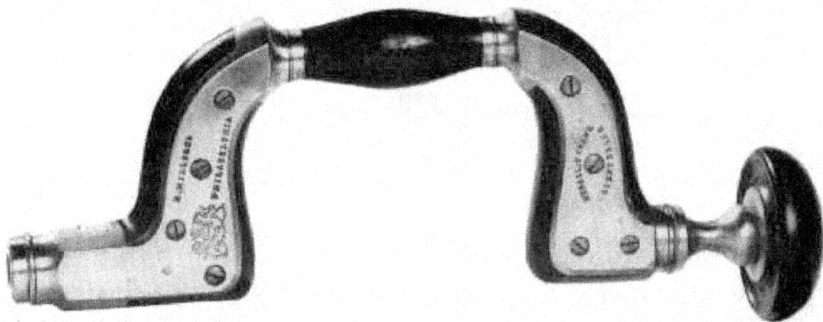

Figure 57.—1876: JAPANNED AND SPLINTED WITH HEAVY BRASS, this brace was among the award-winning tools exhibited at the Centennial by E. Mills and Company of Philadelphia. (Smithsonian photo 49792-D.)

---

[1] Ibid., pp. 14, 44, 5.

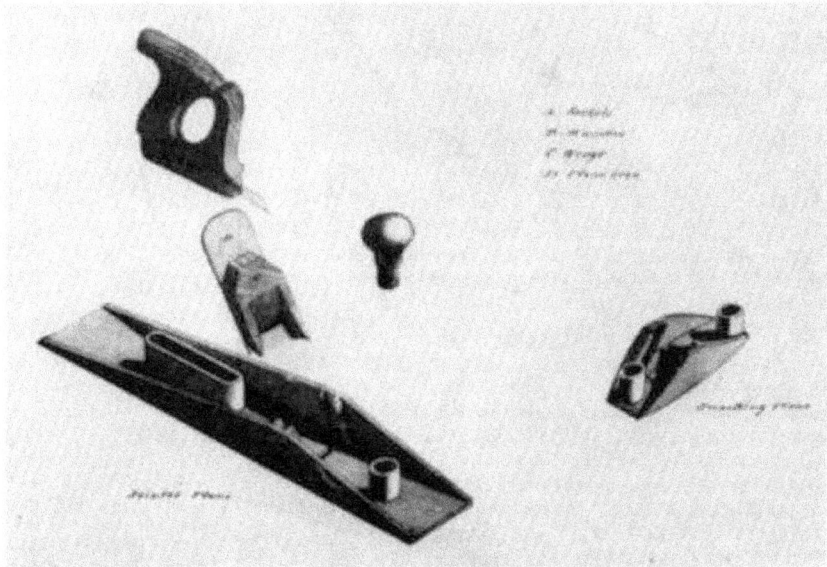

Figure 58.—1827: THE BENCH PLANES exhibited at Philadelphia in 1876 were a radical departure from the traditional. In 1827 H. Knowles patented an iron-bodied bench plane that portended a change in form that would witness a substitution of steel for wood in all critical areas of the tool's construction, and easy adjustment of the cutting edge by a setscrew, and an increased flexibility that allowed one plane to be used for several purposes. (Wash drawing from the restored patent drawings, August 24, 1827, U.S. Patent Office, Record Group 241, the National Archives.)

Figure 59.—1857: THE ADDITION OF METALLIC PARTS to critical areas of wear as suggested by M.B. Tidey did not at first radically alter the design of the bench plane. (Wash drawing from U.S. Patent Office, March 24, 1857, Record Group 241, the National Archives.)

It is the plane, however, that best exemplifies the progress of tool design. In 1876, American planemakers were enthusiastically credited with having achieved "an important change in the structure of the tool."[1] Although change had been suggested by American patentees as early as the 1820's, mass production lagged until after the Civil War, and the use of this new tool form was not widespread outside of the United States. Hazard Knowles of Colchester, Connecticut, in 1827, patented a plane stock of cast iron which in many respects was a proto-type of later Centennial models (fig. 58).[2] It is evident, even in its earliest manifestation, that the quest for improvement of the bench plane did not alter its sound design. In 1857, M.B. Tidey (fig. 59) listed several of the goals that motivated planemakers:

---

[1] Ibid., p. 13.
[2] Restored patent 4,859X, August 24, 1827, National Archives, Washington, D.C.

First to simplify the manufacturing of planes; second to render them more durable; third to retain a uniform mouth; fourth to obviate their clogging; and fifth the retention of the essential part of the plane when the stock is worn out.[1]

By far the greatest number of patents was concerned with perfecting an adjustable plane iron and methods of constructing the sole of a plane so that it would always be "true." Obviously the use of metal rather than the older medium, wood, was a natural step, but in the process of changing from the wood to the iron-bodied bench plane there were many transitional suggestions that combined both materials. Seth Howes of South Chatham, Massachusetts, in U.S. patent 37,694, specified:

This invention relates to an improvement in that class of planes which are commonly termed "bench-planes," comprising the foreplane, smoothing plane, jack plane, jointer, &c.

The invention consists in a novel and improved mode of adjusting the plane-iron to regulate the depth of the cut of the same, in connection with an adjustable cap, all being constructed and arranged in such a manner that the plane-iron may be "set" with the greatest facility and firmly retained in position by the adjustment simply of the cap to the plane-iron, after the latter is set, and the cap also rendered capable of being adjusted to compensate for the wear of the "sole" or face of the plane stock.

The stock of Howes' plane was wood combined with metal plates, caps, and screws. Thomas Worrall of Lowell was issued patent 17,657 for a plane based on the same general principle (fig. 60). Worrall claimed in his specifications of June 23, 1857:

the improved manufacture of [the] carpenter's bench plane or jointer as made with its handle, its wooden stock to which said handle is affixed, and a separate metallic cutter holder, and cutter clamping devices arranged together substantially as specified.

Finally patentees throughout the 19th century, faced with an increasing proliferation of tool types, frequently sought to perfect multipurpose implements of a type best represented later by the ubiquitous Stanley plane. The evolution of the all-purpose idea, which is incidentally not peculiar to hand tools alone, can be seen from random statements selected from U.S. patents for the improvement of bench

---

[1] U.S. pat. 16,889, U.S. Patent Office, Washington, D.C. The numbered specifications that follow may be found in the same place.

planes. In 1864 Stephen Williams in the specifications of his patent 43,360 stated:

I denominate my improvement the "universal smoothing plane," because it belongs to that variety of planes in which the face is made changeable, so that it may be conveniently adapted to the planing of curved as well as straight surfaces. By the use of my improvement surfaces that are convex, concave, or straight may be easily worked, the face of the tool being readily changed from one form to another to suit the surface to which it is to be applied.

The announced object of Theodore Duval's improved grooving plane (pat. 97,177) was "to produce in one tool all that is required to form grooves of several different widths." None was more appealing than Daniel D. Whitker's saw-rabbet plane (pat. 52,478) which combined "an adjustable saw with an adjustable fence or gage, both being attached to a stock with handle similar to a plane, forming together a tool combining the properties of the joiner's plow and fillister" (fig. 61). Nor was Whitker's idea simply a drawing-board exercise. It was produced commercially and was well advertised, as seen in the circular reproduced in figure 62.

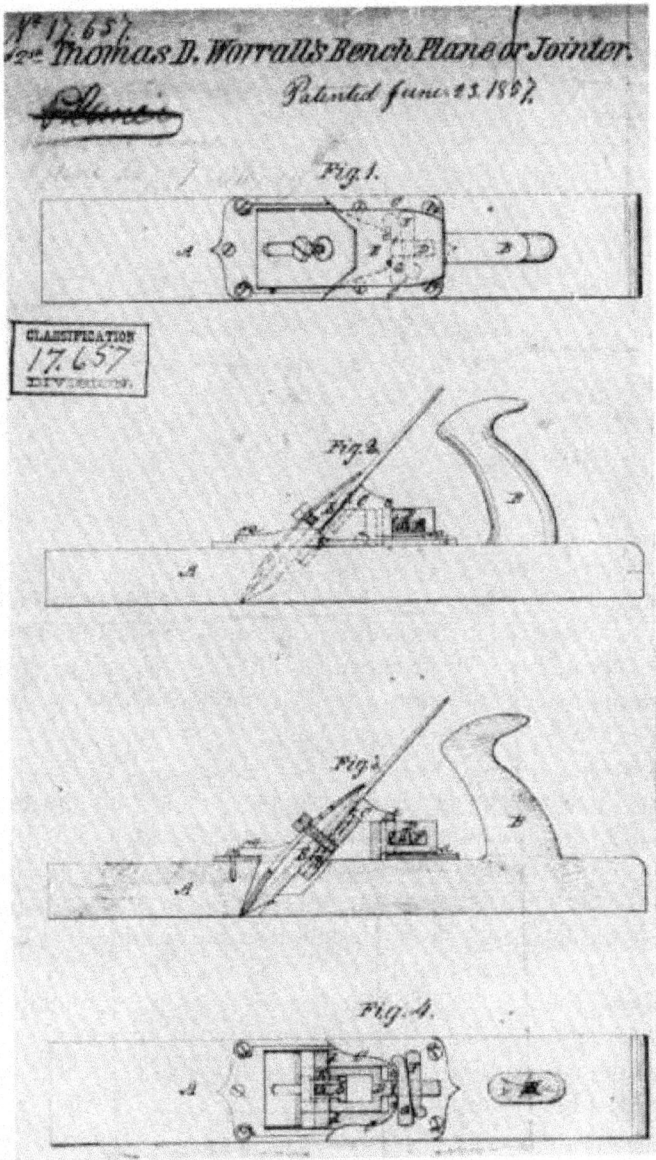

Figure 60.—1857: IN A VARIETY OF ARRANGEMENTS, the addition of metal plates, caps, and screws at the mouth of the plane, as shown in Thomas Worrall's drawing, proved a transitional device that preserved the ancient shape of the tool and slowed the introduction of bench planes made entirely of iron. (Wash drawing from U.S. Patent Office, June 23, 1857, Record Group 241, the National Archives.)

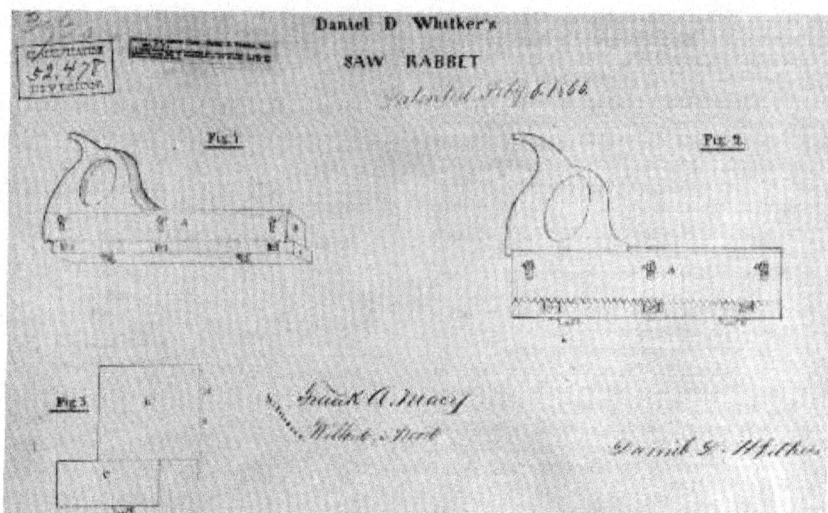

Figure 61.—1865: NOT ALL MULTIPURPOSE INNOVATIONS resulted from the use of new materials. Daniel D. Whitker patented a combination saw and rabbet plane little different from one illustrated by André-Jacob Roubo in his *L'Art du menuisier* in 1769. (Wash drawing from U.S. Patent Office, October 4, 1865, Record Group 241, the National Archives.)

In sum, these ideas produced a major break with the traditional shape of the bench plane. William Foster in 1843 (pat. 3,355), Birdsill Holly in 1852 (pat. 9,094), and W.S. Loughborough in 1859 (pat. 23,928) are particularly good examples of the radical departure from the wooden block. And, in the period after the Civil War, C.G. Miller (discussed on p. 213 and in fig. 63), B.A. Blandin (fig. 64), and Russell Phillips (pat. 106,868) patented multipurpose metallic bench planes of excellent design. It should be pointed out that the patentees mentioned above represent only a few of the great number that tried to improve the plane. Only the trend of change is suggested by the descriptions and illustrations presented here. The cumulative effect awaited a showcase, and the planemakers found it at the Centennial Exhibition of 1876 held in Philadelphia.

# INVISIBLE
# RUBBER WEATHER STRIPS,
## FOR DOORS AND WINDOWS.

(WHITKER'S PATENT.)

Price, No. 1, 6 Cts.

It is entirely out of sight and does not interfere with the free action of either

### DOOR OR WINDOW,

and effectually excludes all Snow, Rain, Wind and Dust.

Price, No. 2, 7 Cts.

It stops all rattle of the Windows and lessens the noise from the street.

It can be applied to any Door or Window without injury, as no nails or screws are used.

Price, No. 3, 8 Cts.

It is the cheapest and most durable STRIP in use, and

### CANNOT WARP OR SPLIT,

as no moulding is used.

A house can be

### Warmed with one-half the Fuel.

Price, Saw, Plane, $1.00
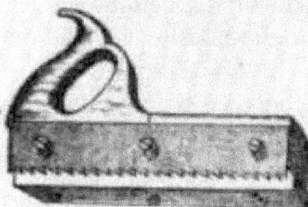

These STRIPS are now in use in many of the best Public Buildings in the country, such as the Astor House,

U. S. Custom House,

U. S. Express Office,

&c., &c.,

Price, Round, $1.00
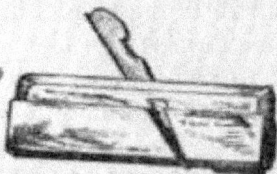

and hundreds of the finest private residences in the city and country, and the universal expression is they

would rather pay two prices for the INVISIBLE in preference to having any of the Moulding Strips as a gift.

### State and County Rights for Sale,

## PRATT & GREEN,
### 208 BROADWAY, Room No. 7.

Chapin & Simpson, Printers, 14 Ann street, N. Y.

Figure 62.—ABOUT 1865: THE PROGRESS OF AN IDEA from an 18th-century encyclopedia through an American patentee to commercial reality can be seen in this flier advertising Whitker's saw-rabbet. (Smithsonian Institution Library. Smithsonian photo 56629.)

The impact of these new planes at the Exhibition caused some retrospection among the judges:

The planes manufactured in Great Britain and in other countries fifty years ago were formed of best beech-wood; the plane irons were of steel and iron welded together; the jointer plane, about 21 inches long, was a bulky tool; the jack and hand planes were of the same materials. Very little change has been made upon the plane in Great Britain, unless in the superior workmanship and higher quality of the plane iron.[1]

The solid wood-block plane, varying from country to country only in the structure of its handles and body decoration, had preserved its integrity of design since the Middle Ages. At the Centennial, however, only a few examples of the old-type plane were exhibited. A new shape dominated the cases. Designated by foreign observers as the American plane, it received extended comment. Here was a tool

constructed with a skeleton iron body, having a curved wooden handle; the plane iron is of the finest cast-steel; the cover is fitted with an ingenious trigger at the top, which, with a screw below the iron, admits of the plane iron being removed for sharpening and setting without the aid of the hammer, and with the greatest ease. The extensive varieties of plane iron in use are fitted for every requirement; a very ingenious arrangement is applied to the tools for planing the insides of circles or other curved works, such as stair-rails, etc. The sole of the plane is formed of a plate of tempered steel about the thickness of a handsaw, according to the length required, and this plate is adapted to the curve, and is securely fixed at each end. With this tool the work is not only done better but in less time than formerly. In some exhibits the face of the plane was made of beech or of other hard wood, secured by screws to the stock, and the tool becomes a hybrid, all other parts remaining the same as in the iron plane.[2]

The popularity of Bailey's patented planes (fig. 65), the type so praised above, was by no means transitory. In 1884 the Boston firm of Goodnow & Wightman, "Importers, Manufacturers and Dealers in Tools of all kinds," illustrated the several planes just described and assured prospective buyers that

These tools meet with universal approbation from the best Mechanics. For beauty of style and finish they are unequalled, and the great convenience in operating renders them the cheapest Planes in use; they are

---

[1] Walker, ed., Reports and Awards, group 15, p. 13.
[2] Ibid.

SELF-ADJUSTING in every respect; and each part being made INTERCHANGEABLE, can be replaced at a trifling expense.[1]

By 1900 an advertisement for Bailey's planes published in the catalogue of another Boston firm, Chandler and Farquhar, indicated that "over 900,000" had already been sold.[2]

Other mass-produced edge tools—axes, adzes, braces and bits, augers, saws, and chisels—illustrated in the trade literature of the toolmakers became, as had the iron-bodied bench plane, standard forms. In the last quarter of the 19th century the tool catalogue replaced Moxon, Duhamel, Diderot, and the builders' manuals as the primary source for the study and identification of hand tools. The Centennial had called attention to the superiority of certain American tools and toolmakers. The result was that until the end of the century, trade literature faithfully drummed the products that had proven such "an attraction to the numerous artisans who visited the Centennial Exhibition from the United States and other countries."[3]

---

[1] Tools (Boston, 1884), p. 54 [in the Smithsonian Institution Library].
[2] Tools and Supplies (June 1900), no. 85 [in the Smithsonian Institution Library].
[3] Walker, op. cit. (footnote 19), p. 14.

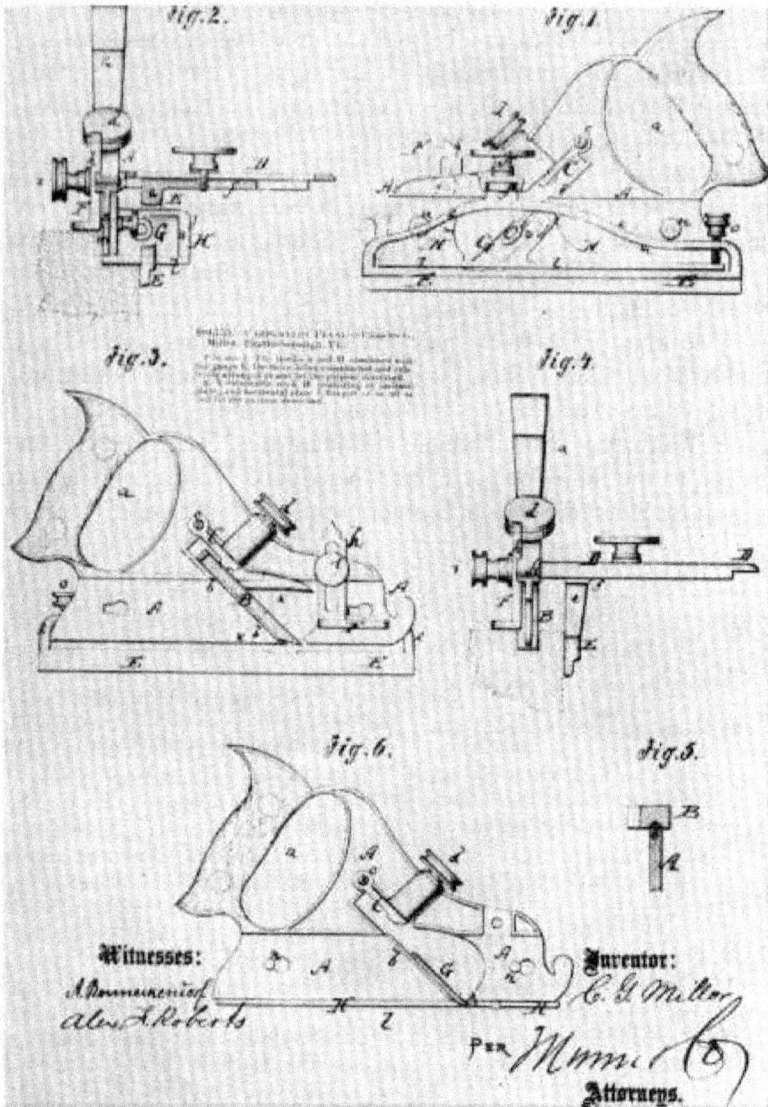

Figure 63.—1870: THE METALLIC VERSION OF THE PLOW PLANE later produced by Stanley and Company was patented by [Charles] G. Miller as a tool readily "convertible into a grooving, rabbeting, or smoothing plane." In production this multipurpose plow gained an elaborate decoration (fig. 51) nowhere suggested in Miller's specification. (Wash drawing from U.S. Patent Office, June 28, 1870, Record Group 241, the National Archives.)

Peter C. Welsh

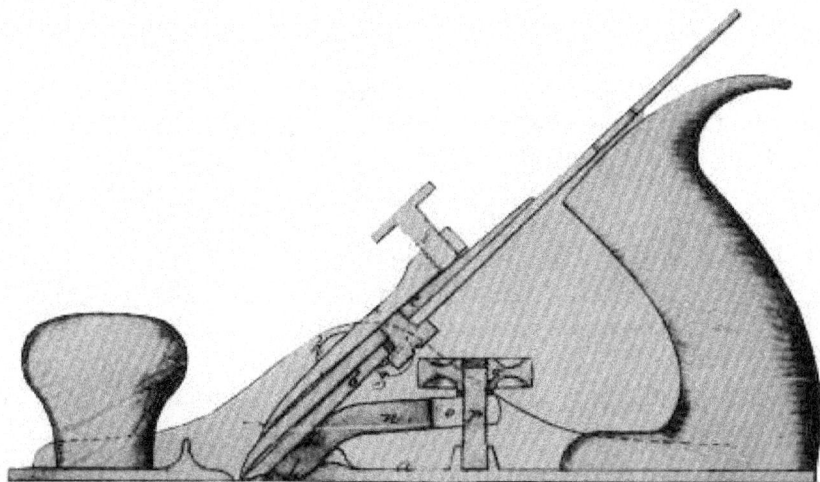

Figure 64.—1867: THE DRAWING accompanying B.A. Blandin's specification for an "Improvement in Bench Planes" retained only the
familiarly shaped handle or tote of the traditional wood-bodied plane.
This new shape rapidly became the standard form of the tool with later
variations chiefly related to the adjustability of the plane-iron and sole.
(Wash drawing from U.S. Patent Office, May 7, 1867, Record Group
241, the National Archives.)

Collins and Company of New York City had been given commendation for the excellence of their axes; through the end of the century,
Collins' brand felling axes, broad axes, and adzes were standard items,
as witness Hammacher, Schlemmer and Company's catalogue of
1896.[1] Disston saws were a byword, and the impact of their exhibit at
Philadelphia was still strong, as judged from Baldwin, Robbins' catalogue of 1894. Highly recommended was the Disston no. 76, the
"Centennial" handsaw with its "skew back" and "apple handle." Jennings' patented auger bits were likewise standard fare in nearly every
tool catalogue.[2] So were bench planes manufactured by companies that
had been cited at Philadelphia for the excellence of their product;
namely, The Metallic Plane Company, Auburn, New York; The Mid-

[1] Tools for All Trades (New York, 1896), item 75 [in the Smithsonian Institution Library].
[2] See Baldwin, Robbins & Co.: Illustrated Catalogue (Boston, 1894), pp. 954,
993 [in the Smithsonian Institution Library].

dletown Tool Company, Middletown, Connecticut; Bailey, Leonard, and Company, Hartford; and The Sandusky Tool Company, Sandusky, Ohio.[1]

An excellent indication of the persistence of the Centennial influence, and of the tool catalogue as source material, is seen in Chandler and Farquhar's illustrated pamphlet of 1900. Their advertisement for Barber's improved ratchet brace (fig. 66), a tool much admired by the Centennial judges, amply illustrates the evolution of design of a basic implement present in American society since the first years of settlement. The Barber brace represents the ultimate sophistication of a tool, achieved through an expanded industrial technology rather than by an extended or newly found use for the device itself. It is a prime example of the transition of a tool from Moxon to its perfected form in the 20th century:

These Braces possess the following points of superiority: The Sweep is made from Steel; the Jaws are forged from Steel; the Wood Handle has brass rings inserted in each end so it cannot split off; the Chuck has a hardened Steel antifriction washer between the two sockets, thus reducing the wear. The Head has a bearing of steel balls, running on hard steel plates, so no wear can take place, as the friction is reduced to the minimum. The Brace is heavily nickel-plated and warranted in every particular. We endeavor to make these goods as nearly perfection as is possible in durability, quality of material and workmanship, and fineness and beauty of finish.[2]

---

[1] Walker, op. cit. (footnote 19), p. 14.
[2] Tools and Supplies, op. cit. (footnote 22).

CHANDLER & FARQUHAR, BOSTON, MASS.

## Bailey's Patent Adjustable Planes.

**OVER 900,000 ALREADY SOLD.**

### Iron Planes.

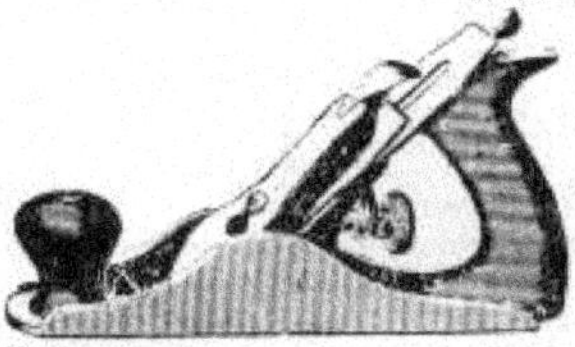

These Planes meet with universal approbation from the best Mechanics, as their extensive sale abundantly testifies. For beauty of style and finish, they are unequaled, and the superior methods for adjusting them readily in all their parts, render them economical to the owner.

| No. | | | Each. |
|---|---|---|---|
| 1. | Smooth Plane, | 5½ inches in Length, 1¼ inch Cutter............ | $1.13 |
| 2. | Smooth Plane, | 7 inches in Length, 1⅝ inch Cutter............ | 1.38 |
| 3. | Smooth Plane, | 8 inches in Length, 1¾ inch Cutter............ | 1.50 |
| 4. | Smooth Plane, | 9 inches in Length, 2 inch Cutter............ | 1.62 |
| 4½. | Smooth Plane, | 10 inches in Length, 2⅜ inch Cutter............ | 1.87 |
| 5. | Jack Plane, | 14 inches in Length, 2 inch Cutter............ | 1.87 |
| 6. | Fore Plane, | 18 inches in Length, 2⅜ inch Cutter............ | 2.13 |
| 7. | Jointer Plane, | 22 inches in Length, 2⅜ inch Cutter............ | 2.75 |
| 8. | Jointer Plane, | 24 inches in Length, 2⅝ inch Cutter............ | 3.25 |
| 0. | Block Plane, | 10 inches in Length, 2 inch Cutter............ | 3.00 |

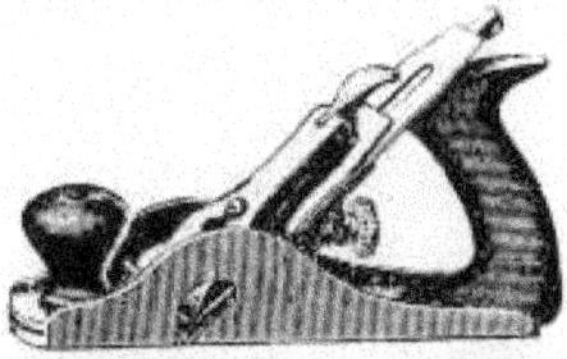

| | | |
|---|---|---|
| 10½. | Carriage Makers' Rabbet Plane, 9 inches Length, 2⅛ inch Cutter. | 1.87 |
| 10. | Carriage Makers' Rabbet Plane, 14 inches Length, 2⅛ inch Cutter. | 2.25 |
| 11. | Belt Makers' Plane, 2⅜ inch Cutter............ | 1.50 |

### Bailey's Adjustable Circular Plane.

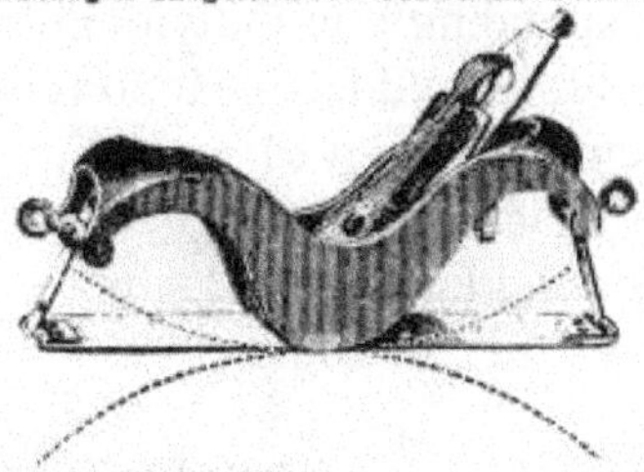

No. 13. Circular Plane, 1¾ inch Cutter............ 2.00

This Plane has a Flexible Steel Face, and by means of the thumb screws at each end of the Stock, can be easily adapted to plane circular work—either concave or convex.

Figure 65.—1900: AMERICAN PLANEMAKERS had been cited at the Philadelphia Centennial as having introduced a dramatic change in the nature of the tool. Although wood-bodied planes continued to be used, they were outdated and in fact anachronistic by the close of the 19th century. From the 1870's forward, it was the iron-bodied plane, most frequently Bailey's, that enlivened the trade literature. (Catalogue of Chandler and Farquhar, Boston, 1900. Smithsonian photo 55798.)

## BARBER IMPROVED RATCHET BRACES.

These Braces possess the following points of superiority: The Sweep is made from Steel; the Jaws are forged from Steel; the Wood Handle has brass rings inserted in each end so it cannot split off; the Chuck has a hardened Steel anti-friction washer between the two sockets, thus reducing the wear. The Head has a bearing of steel balls, running on hardened steel plates, so no wear can take place, as the friction is reduced to the minimum. The Brace is heavily nickel-plated and warranted in every particular. We endeavor to make these goods as nearly perfection as is possible in durability, quality of material and workmanship, and fineness and beauty of finish.

| Nos, | | 33 | 38 | 31 |
| Sweep, inches, | | 8 | 10 | 12 |
| Price, each, | | $1.50 | 2.00 | 2.10 |

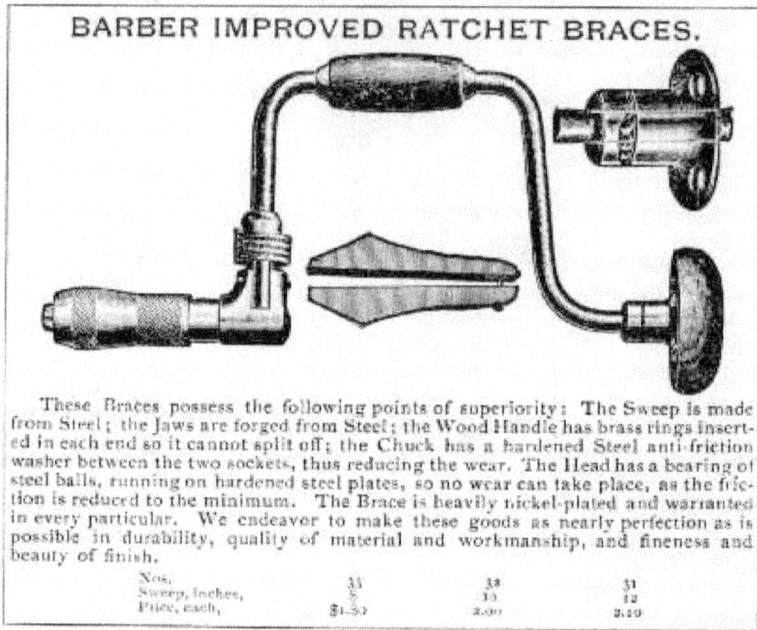

Figure 66.—1900: FEW TOOLS SUGGEST MORE CLEARLY the influence of modern industrial society upon the design and construction of traditional implements than Barber's ratchet brace. It is not without interest that as the tools of the wood craftsman became crisply efficient, his work declined correspondingly in individuality and character. The brace and the plane, as followed from Moxon through the trade literature of the late 19th century, achieved perfection in form and operation at a time when their basic functions had been usurped by machines. (Catalogue of Chandler and Farquhar, Boston, 1900. Smithsonian photo 56626.)

The description of Barber's brace documents a major technical change: wood to steel, leather washers to ball bearings, and natural patina to nickel plate. It is also an explanation for the appearance and shape of craftmen's tools, either hand forged or mass produced. In each case, the sought-after result in the form of a finished product has been an implement of "fineness and beauty." This quest motivated three centuries of toolmakers and brought vitality to hand-tool design. Moxon had advised:

76

He that will a good Edge win,
Must Forge thick and Grind thin.[1]

If heeded, the result would be an edge tool that assured its owner "ease and delight."[2] Throughout the period considered here, the most praiseworthy remarks made about edge tools were variations of either "unsurpassed in quality, finish, and beauty of style" or, more simply, commendation for "excellent design and superior workmanship."[3] The hand tool thus provoked the same value words in the 19th as in the 17th century.

The aesthetics of industrial art, whether propounded by Moxon or by an official at the Philadelphia Centennial, proved the standard measure by which quality could be judged. Today these values are particularly valid when applied to a class of artifacts that changed slowly and have as their prime characteristics anonymity of maker and date. With such objects the origin, transition, and variation of shape are of primary interest. Consider the common auger whose "Office" Moxon declared "is to make great round holes" and whose importance was so clearly stressed at Philadelphia in 1876.[4] Neither its purpose nor its gross appearance (a T-handled boring tool) had changed. The tool did, however, develop qualitatively through 200 years, from a pod or shell to a spiral bit, from a blunt to a gimlet point, and from a hand-fashioned to a geometrically exact, factory-made implement: innovations associated with Cooke (1770), L'Hommedieu (1809), and Jennings (1850's). In each instance the tool was improved—a double spiral facilitated the discharge of shavings, a gimlet point allowed the direct insertion of the auger, and machine precision brought mathematical accuracy to the degree of twist. Still, overall appearance did not change. At the Centennial, Moxon would have recognized an auger, and, further, his lecture on its uses would have been singularly current. The large-bore spiral auger still denoted a mortise, tenon, and trenail mode of building in a wood-based technology; at the same time its near cousin, the wheelwright's reamer, suggested the reliance upon a transport dependent upon wooden hubs. The auger in its perfected form—fine steel, perfectly machined, and highly finished—contrasted with an auger of earlier vintage will clearly show the advance from

---

[1] Mechanick Exercise ..., p. 62.
[2] Ibid., p. 95.
[3] Walker, op. cit. (footnote 19), pp. 31–49.
[4] Mechanick Exercises ..., p. 94.

forge to factory, but will indicate little new in its method of use or its intended purpose.

Persons neither skilled in the use of tools nor interested in technical history will find that there is another response to the common auger, as there was to the upholsterer's hammer, the 18th-century brace, or the saw with the custom-fitted grip. This is a subjective reaction to a pleasing form. It is the same reaction that prompted artists to use tools as vehicles to help convey lessons in perspective, a frequent practice in 19th-century art manuals. The harmony of related parts—the balance of shaft and handle or the geometry of the twist—makes the auger a decorative object. This is not to say that the ancient woodworker's tool is not a document attesting a society's technical proficiency—ingenuity, craftsmanship, and productivity. It is only to suggest again that it is something more; a survival of the past whose intrinsic qualities permit it to stand alone as a bridge between the craftsman's hand and his work; an object of considerable appeal in which integrity of line and form is not dimmed by the skill of the user nor by the quality of the object produced by it.

In America, this integrity of design is derived from three centuries of experience: one of heterogeneous character, the mid-17th to the mid-18th; one of predominately English influence, from 1750 to 1850; and one that saw the perfection of basic tools, by native innovators, between 1850 and the early 20th century. In the two earlier periods, the woodworking tool and the products it finished had a natural affinity owing largely to the harmony of line that both the tool and finished product shared. The later period, however, presents a striking contrast. Hand-tool design, with few exceptions, continued vigorous and functional amidst the confusion of an eclectic architecture, a flurry of rival styles, the horrors of the jigsaw, and the excesses of Victorian taste. In conclusion, it would seem that whether seeking some continuous thread in the evolution of a national style, or whether appraising American contributions to technology, such a search must rest, at least in part, upon the character and quality of the hand tools the society has made and used, because they offer a continuity largely unknown to other classes of material survivals.

# BIBLIOGRAPHY

*Book of trades, or library of the useful arts.* 1st Amer. ed. Whitehall, N.Y., 1807.

*Boy's book of trades.* London, 1866.

*The cabinetmaker in eighteenth-century Williamsburg.* (Williamsburg Craft Series.)

Williamsburg, Va., 1963.

COMENIUS, JOHANN AMOS. *Orbis sensualium pictus.* Transl. Charles Hoole. London,

1664, 1685, 1777, et al.

COTTER, JOHN L. *Archeological excavations at Jamestown, Virginia.* (No. 4 in Archeological

Research Series.) Washington: National Park Service, 1958.

DIDEROT, DENIS. *L'encyclopédie, ou dictionnaire raisonné des sciences, des arts et des métiers.*

Paris, 1751–1765.

EARLY AMERICAN INDUSTRIES ASSOCIATION. *Chronicle.* Williamsburg, Va., 1933+.

GILLISPIE, CHARLES COULSTON, ed. *A Diderot pictorial encyclopedia of trades and industry.*

New York: Dover Publications, Inc., 1959.

GOODMAN, W.L. *The history of woodworking tools.* London: G. Bell and Sons, Ltd.,

1964.

HOLTZAPFFEL, CHARLES. *Turning and mechanical manipulations.* London [1846].

KNIGHT, EDWARD HENRY. *Knight's American mechanical dictionary.* New York, 1874–1876.

MARTIN, THOMAS. *The circle of the mechanical arts.* London, 1813.

MERCER, HENRY C. *Ancient carpenters' tools.* Doylestown, Pennsylvania: The Bucks

County Historical Society, 1951.

MOXON, JOSEPH. *Mechanick exercises.* 3rd ed. London, 1703.

NICHOLSON, PETER. *The mechanic's companion.* Philadelphia, 1832.

PETERSEN, EUGENE T. *Gentlemen on the frontier: A pictorial record of the culture of Michilimackinac.*

Mackinac Island, Mich., 1964.

PETRIE, SIR WILLIAM MATTHEW FLINDERS. *Tools and weapons illustrated by the Egyptian*

*collection in University College, London.* London, 1917.

ROUBO, ANDRÉ-JACOB. *L'art du menuisier.* (In Henri-Louis Duhamel du Monceau,

*Descriptions des arts et métiers.*) Paris, 1761–1788.

SACHS, HANS. *Das Ständebuch: 114 Holzschnitte von Jost Amman, mit Reimen von Hans*

*Sachs.* Leipzig: Insel-Verlag, 1934.

SINGER, CHARLES, et al. *A history of technology.* 5 vols. New York and London:

Oxford University Press, 1954–1958.

SLOANE, ERIC. *A museum of early American tools.* New York: Wilfred Funk, Inc.,

1964.

TOMLINSON, CHARLES. *Illustrations of trades.* 2nd ed. London, 1867.

WELSH, PETER C. "The Decorative Appeal of Hand Tools," *Antiques*, vol. 87, no. 2,

February 1965, pp. 204–207.

---- U.S. patents, 1790–1870: New uses for old ideas. Paper 48 in *Contributions*

*from the Museum of History and Technology:* Papers 45–53 (U.S. National

Museum Bulletin 241), by various authors; Washington: Smithsonian Institution,

1965.

WILDUNG, FRANK H. *Woodworking tools at Shelburne Museum.* (No. 3 in Museum

Pamphlet Series.) Shelburne, Vermont: The Shelburne Museum, 1957.

*Paper 51, pages 178–228, from* UNITED STATES NATIONAL MUSEUM BULLETIN

CONTRIBUTIONS FROM
THE MUSEUM OF HISTORY
AND TECHNOLOGY
SMITHSONIAN INSTITUTION
WASHINGTON, D.C.

**Iron Making in the Olden Times as instanced in the Ancient Mines, Forges, and Furnaces of The Forest of Dean**
**H. G. Nicholls**
Benediction Classics, 2011
82 pages
ISBN: 978-1-78139-007-8

Available from www.amazon.com,
www.amazon.co.uk

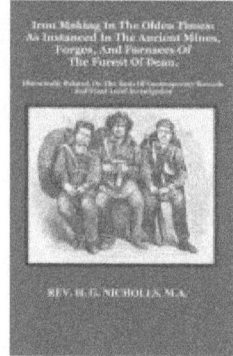

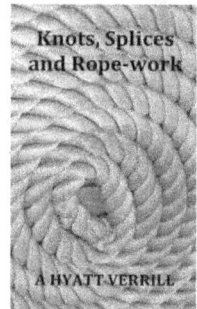

In this fully illustrated book, the Rev. H. G. Nicholls, M.A., studies the historical evidence of the Iron Ore Mining in the Forest of Dean from the earliest times. The book contains these words in the introduction: In the year 1780, wrote Mr. Wyrrall, in his valuable MS. on the ancient iron works of the Forest:- "There are, deep in the earth, vast caverns scooped out by men's hands, and large as the aisles of churches; and on its surface are extensive labyrinths worked among the rocks, and now long since overgrown with woods, which whosoever traces them must see with astonishment, and incline to think them to have been the work of armies rather than of private labourers. They certainly were the toil of many centuries, and this perhaps before they thought of searching in the bowels of the earth for their ore-whither, however, they at length naturally pursued the veins, as they found them to be exhausted near the surface."

**Knots, Splices and Rope-work (fully illustrated)**
**A. Hyatt Verrill**
Oxford City Press, 2011
108 pages
ISBN: 978-1-78139-012-2

Available from www.amazon.com,
www.amazon.co.uk

This book offers easy to follow directions for making all the useful and ornamental knots that are in common use. There are chapters covering splicing, pointing, seizing and serving, etc. This text is ideal for campers, yachtsmen, Boy Scouts, and anyone interested in knots.

Italian Harpsichord-Building in the 16th and
17th Centuries - Fully illustrated
**John D. Shortridge**
Oxford City Press, 2011
38 pages
ISBN: 978-1-78139-000-9

Available from www.amazon.com,
www.amazon.co.uk

This edition comes fully illustrated. From
the introduction to the book: "The making of
harpsichords flourished in Italy throughout the 16th and 17th
centuries. The Italian instruments were of simpler construction
than those built by the North Europeans, and they lacked the
familiar second manual and array of stops. In this paper, typical
examples of Italian harpsichords from the Hugo Worch Collection
in the United States National Museum are described in detail and
illustrated. Also, the author offers an explanation for certain
puzzling variations in keyboard ranges and vibrating lengths of
strings of the Italian harpsichords. The Author: John D. Shortridge
is associate curator of cultural history in the United States
National Museum, Smithsonian Institution."

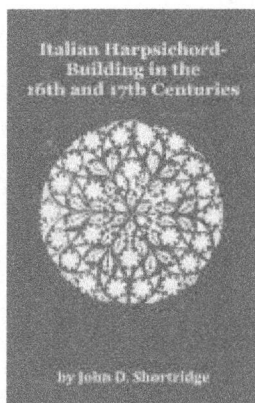

**Pen Drawing, An Illustrated Treatise**
**Charles D Maginnis**
Oxford City Press, 2011
136 pages
ISBN: 978-1-78139-013-9

Available from www.amazon.com,
www.amazon.co.uk

This is a relatively short book, published in
1899, showing some exquisite examples of pen
drawing, and instructing the reader on how to produce them.
Maginnis was a celebrated architect, and his love of buildings
comes through in the text. The tone of the book can be a bit dated,
but that adds to its historical interest, and the instruction is as true
today as it was when the text was written.

Also from Benediction Books …
**Wandering Between Two Worlds: Essays on Faith and Art**
**Anita Mathias**
Benediction Books, 2007
152 pages
ISBN: 0955373700

Available from www.amazon.com, www.amazon.co.uk

In these wide-ranging lyrical essays, Anita Mathias writes, in lush, lovely prose, of her naughty Catholic childhood in Jamshedpur, India; her large, eccentric family in Mangalore, a sea-coast town converted by the Portuguese in the sixteenth century; her rebellion and atheism as a teenager in her Himalayan boarding school, run by German missionary nuns, St. Mary's Convent, Nainital; and her abrupt religious conversion after which she entered Mother Teresa's convent in Calcutta as a novice. Later rich, elegant essays explore the dualities of her life as a writer, mother, and Christian in the United States-- Domesticity and Art, Writing and Prayer, and the experience of being "an alien and stranger" as an immigrant in America, sensing the need for roots.

**About the Author**

Anita Mathias is the author of *Wandering Between Two Worlds: Essays on Faith and Art.* She has a B.A. and M.A. in English from Somerville College, Oxford University, and an M.A. in Creative Writing from the Ohio State University, USA. Anita won a National Endowment of the Arts fellowship in Creative Nonfiction in 1997. She lives in Oxford, England with her husband, Roy, and her daughters, Zoe and Irene.

Anita's website:
     http://www.anitamathias.com, and
Anita's blog Dreaming Beneath the Spires:
     http://dreamingbeneaththespires.blogspot.com

**The Church That Had Too Much**
**Anita Mathias**
Benediction Books, 2010
52 pages
ISBN: 9781849026567

Available from www.amazon.com, www.amazon.co.uk

The Church That Had Too Much was very well-intentioned. She
wanted to love God, she wanted to love people, but she was both ham-
pered by her muchness and the abundance of her possessions, and
beset by ambition, power struggles and snobbery. Read about the sur-
prising way The Church That Had Too Much began to resolve her
problems in this deceptively simple and enchanting fable.


**About the Author**

Anita Mathias is the author of *Wandering Between Two Worlds: Essays on Faith and Art*. She has a B.A. and M.A. in English from
Somerville College, Oxford University, and an M.A. in Creative Writing from the Ohio State University, USA. Anita won a National
Endowment of the Arts fellowship in Creative Nonfiction in 1997.
She lives in Oxford, England with her husband, Roy, and her daughters, Zoe and Irene.

Anita's website:
    http://www.anitamathias.com, and
Anita's blog Dreaming Beneath the Spires:
    http://dreamingbeneaththespires.blogspot.com

www.ingramcontent.com/pod-product-compliance
Lightning Source LLC
Chambersburg PA
CBHW021205020426
42331CB00003B/212